T0142020

Studies in Fuzziness and Soft Computing

Volume 311

Series editor

Janusz Kacprzyk, Polish Academy of Sciences, Warsaw, Poland
e-mail: kacprzyk@ibspan.waw.pl

For further volumes:
http://www.springer.com/series/2941

About this Series

The series "Studies in Fuzziness and Soft Computing" contains publications on various topics in the area of soft computing, which include fuzzy sets, rough sets, neural networks, evolutionary computation, probabilistic and evidential reasoning, multi-valued logic, and related fields. The publications within "Studies in Fuzziness and Soft Computing" are primarily monographs and edited volumes. They cover significant recent developments in the field, both of a foundational and applicable character. An important feature of the series is its short publication time and world-wide distribution. This permits a rapid and broad dissemination of research results.

Li Yan · Zongmin Ma · Fu Zhang

Fuzzy XML Data Management

Li Yan
School of Software
Northeastern University
Shenyang
People's Republic of China

Zongmin Ma
College of Information Science and Engineering
Northeastern University
Shenyang
People's Republic of China

Fu Zhang
College of Information Science and Engineering
Northeastern University
Shenyang
People's Republic of China

ISSN 1434-9922 ISSN 1860-0808 (electronic)
ISBN 978-3-662-50927-2 ISBN 978-3-642-44899-7 (eBook)
DOI 10.1007/978-3-642-44899-7
Springer Heidelberg New York Dordrecht London

Softcover reprint of the hardcover 1st edition 2014

Printed on acid-free paper

Springer is part of Springer Science+Business Media (www.springer.com)

Preface

The Web is a huge information resource depository all around the world and the huge amount of information on the Web is getting larger and larger every day. Nowadays the Web is the most important means for people to acquire and publish information. Against this context, it has become very crucial to exchange and share data on the Web. Being the de facto standard for data representation and exchange over the Web, Extensible Markup Language (XML) has emerged and extensively applied in many business, service, and multimedia applications. As a result, a large volume of data is managed today directly in XML format.

XML and related standards allow the easy development of applications that exchange data over the Web. This creates a new set of data management requirements involving XML, such as the needs of constructing, storing, querying, reasoning, and integrating XML documents. For the purpose of XML data management, it is necessary to integrate XML and databases. Various database models, including relational, object-oriented, and object-relational database models as well as conceptual data models, have been used for mapping to and from XML model so that XML model can be extracted from the database models or/and reengineered into the database models. Through reengineering XML model into the relational or object-oriented database model, XML documents can be stored in databases for XML data processing based on database technologies. Through extracting XML model from the relational or object-oriented database model, XML documents can automatically be constructed from databases for database data exchange and share on the Web. Note that XML lacks sufficient power in modeling real-world data and their complex interrelationships in semantics, and the conceptual data models with powerful data abstraction contain clear and rich semantics and do not have data type limitation. Through reengineering XML model into the conceptual data models, XML document integration can be carried out based on the conceptual data models. Through extracting XML model from the conceptual data models, XML documents can conceptually be designed with the conceptual data models.

With the wide and in-depth utilization of XML in diverse application domains, some particularities of data management in concrete applications emerge, which challenge current XML technology. In data- and knowledge-intensive applications, one of the challenges can be generalized as the need to handle imprecise and uncertain information in XML data management. Imprecise and uncertain data can be found, for example, in the integration of data sources and data generation with nontraditional means (e.g., automatic information extraction and data acquirement by sensor and RFID). So it is crucial for Web-based intelligent information systems to explicitly represent and process imprecise and uncertain XML data.

Fuzzy logic has been applied in a large number and in a wide variety of applications and has been a crucial means of implementing machine intelligence. So, in order to bridge the gap between human-understandable soft logic and machine-readable hard logic, fuzzy logic cannot be ignored because none of the usual logical requirements can be guaranteed: there is no centrally defined format for data, no guarantee of truth for assertions made, and no guarantee for consistency. Fuzzy logic has been introduced into databases for fuzzy data management. It can be believed that fuzzy logic can play an important and positive role in XML data management. Currently the researches of fuzzy logic in XML data management are attracting increased attention.

This book goes to great depth concerning the fast growing topic of technologies and approaches of fuzzy XML data management. The topics of this book include representation of fuzzy XML, query of fuzzy XML, fuzzy database models, extraction of fuzzy XML from fuzzy database models, reengineering of fuzzy XML into fuzzy database models, and reasoning of fuzzy XML. Concerning the representation of fuzzy XML, the fuzziness in XML documents, fuzzy XML representation model, and fuzzy XML algebraic operations are discussed. Concerning the query of fuzzy XML, querying fuzzy XML with AND, OR, and NOT predicates is proposed, respectively, and building the index on fuzzy XML query is investigated. Concerning the fuzzy database models, three kinds of fuzzy database models are introduced, which are the fuzzy UML data models, fuzzy relational database models, and fuzzy object-oriented database models. Concerning the extraction of fuzzy XML, extracting fuzzy XML from the fuzzy UML data models, fuzzy relational database models, and fuzzy object-oriented database models is proposed, respectively. Concerning the reengineering of fuzzy XML, reengineering fuzzy XML into the fuzzy UML data models, fuzzy relational database models, and fuzzy object-oriented database models is presented, respectively. Concerning the reasoning of fuzzy XML, reasoning on fuzzy XML with fuzzy Description Logic and fuzzy ontology are investigated.

This book aims to provide a single record of current research in the fuzzy data management with XML. The objective of the book is to provide the state-of-the-art information to researchers, practitioners, and graduate students of the Web intelligence and at the same time serve the data and knowledge engineering professional faced with nontraditional applications that make the application of conventional approaches difficult or impossible. Researchers, graduate students, and information technology professionals interested in XML and fuzzy data processing will find this book a starting point and a reference for their study, research, and development.

We would like to acknowledge all of the researchers in the area of databases, XML, and fuzzy databases. Based on both their publications and the many discussions with some of them, their influence on this book is profound. The materials in this book are the outgrowth of research conducted by the authors in recent years. The initial research work was supported by the *National Natural Science Foundation of China* (60873010, 61370075, 61073139, 61202260), and in part by the *Program for New Century Excellent Talents in University* (NCET-05-0288). We are grateful for the financial support from the *National Natural Science Foundation of China* and the *Ministry of Education of China* through research grant funds. Additionally, the assistance and facilities of Northeastern University, China, are deemed important and highly appreciated. Special thanks go to Janusz Kacprzyk, the series editor of Studies in Fuzziness and Soft Computing, and Thomas Ditzinger, the Senior Editor of Applied Sciences and Engineering of Springer-Verlag, for their advice and help to propose, prepare, and publish this book. This book would not be completed without the support from them.

Shenyang, September 2013

Li Yan
Zongmin Ma
Fu Zhang

Contents

Chapter 1
Databases and XML for Data Management

Abstract A large number of data appears in various real-world application domains, and how to manage the data is particularly important. Databases are created to operate large quantities of data by inputting, storing, retrieving, and managing that data. Over the years, various database models, including conceptual data models (e.g., entity-relationship (ER) model, enhanced entity-relationship (EER) model, and UML data model) and logical database models (e.g., relational database model and object-oriented database model), are developed for information modeling and data management. Moreover, with the popularity of Web-based applications, the requirement for data management has been put on the exchange and share of data over the Web. The eXtensiable Markup Language (XML) provides a Web friendly and well-understood syntax for the exchange of data and impacts on data definition and share on Web. This is creating a new set of data management requirements involving XML. Currently, databases and XML play important roles for data management and have become the main means to realize the data management. In this chapter, databases and XML techniques for data management will be introduced.

1.1 Introduction

With the emergency of a large number of data, the concept of "Data Management" accordingly arose in the 1980s as technology moved from sequential processing (first cards, then tape) to random access processing. As applications moved more and more into real-time, interactive applications, it became particularly important to manage data so that the data would be used in applications conveniently and effectively.

The evolution of databases was initially driven by the requirements of data processing. A database is an organized collection of data. The data is typically organized to model relevant aspects of reality, in a way that supports processes requiring this information. Databases have become the key to data management.

L. Yan et al., *Fuzzy XML Data Management*, Studies in Fuzziness and Soft Computing 311, DOI: 10.1007/978-3-642-44899-7_1,

Information modeling in databases can be carried out at two different levels: *conceptual data modeling* and *logical database modeling*. Basically, the conceptual data models are used for information modeling at a high level of abstraction and at the level of data manipulation, i.e., a low level of abstraction. The entity-relationship (ER) conceptual data model, which was proposed by Chen (1976), has played a crucial role in database design and information systems analysis. Further, in order to overcome its incapability of ER modeling complex objects and semantic relationships. Several new conceptual data models (e.g., the enhanced entity-relationship (EER) model and UML model) are developed. The logical database model is used for information modeling. Database modeling generally starts from the conceptual data models and then the developed conceptual data models are mapped into the logical database models. A logical database model determines the logical structure of a database and fundamentally determines in which manner data can be stored, organized, and manipulated. A database management system (DBMS) is a suite of computer software providing the interface between users and a database or databases.

The development of database technology can be divided into several eras: *navigational*, *relational*, and *post-relational*. Hierarchical and network database models were adopted by DBMSs as *navigational* database models in the 1960s and 1970s. The hierarchical and network database models have the drawbacks that they couple with the need for a formally based database model, which clearly separate the physical and logical model. Therefore, for enhancing the hierarchical and network database models, relational database model is hereby developed. The *relational* database model put forward by Codd (1970), has a simple structure and a solid mathematical foundation. It is made up of ledger-style tables, each used for a different type of entity. It rapidly replaced the hierarchical and network database models and became the dominant database model for commercial database systems. Although there has been great success in using the relational databases for transaction processing, with the breadth and depth of database uses in many emerging areas as diverse as biology and genetics, artificial intelligence, and geographical information systems, it was realized that the relational database model, had semantic and structured drawbacks when it came to modeling of such specialized applications. For example, the data type is very restricted and the data semantics is not rich in the relational database model. In this case, several non-traditional database models (Abiteboul et al. 1995; Elmasri and Navathe 1994; Kim and Lochovsky 1989; Abiteboul and Hull 1987; Hammer and McLeod 1981) were developed in succession to enlarge the application area of databases since the end of the 1970s, and they are called *post-relational database models*. In particular, object-oriented database model is a popular one of the post-relational database models. In object-oriented database models, all real-world entities can be simulated as objects, objects with the same attributes, methods and constraints can be incorporated into classes, and the classes form class hierarchical structures.

Moreover, currently much data is unstructured and is in the form of Web documents, and thus does not fit well into databases such as relational or object-oriented databases. In particular, with the prompt development of Web, the requirement of managing information based on the Web has attracted much attention both from academia and industry. The eXtensible Markup Language (XML) is emerging and gradually accepted as the de facto standard for data description and exchange between various systems and databases over the Internet. This is creating a new set of data management requirements involving XML, such as the need to store and query XML documents. Also, lots of translation techniques have been devised to publish large amounts of existing data in databases (such as g relational and object-oriented databases) in XML format.

Based on the observations above, databases and XML have become the key means to realize the data management. In this chapter, we discuss some common database models, including the logical database models (relational database model and object-oriented database model) and the conceptual data models (ER/EER model and UML data model). Also the XML model will be introduced.

1.2 Logical Database Models

The evolution of database systems was initially driven by the requirements of traditional data processing. Hierarchical and network data models were adopted by database management systems (DBMS) as database models in the 1960s and 1970s. The hierarchical and network data models have the drawbacks that the data models couple with the need for a formally based database model, which clearly separate the physical and logical model. Relational database model, put forward by Codd (1970), has a simple structure and a solid mathematical foundation. It rapidly replaced the hierarchical and network database models and became the dominant database model for commercial database systems.

With the breadth and depth of database uses in many emerging areas as diverse as biology and genetics, artificial intelligence, computer aided design, and geographical information systems, it was realized that the relational database model as defined by Codd, had semantic and structured drawbacks when it came to modeling of such specialized applications. The next evolution of database models took the form of rich data models such as the object-oriented data model (Abiteboul et al. 1995; Elmasri and Navathe 1994; Kim and Lochovsky 1989) and the semantic data models (Abiteboul and Hull 1987; Elmasri and Navathe 1994; Hammer and McLeod 1981).

Relational database model and object-oriented database model are typical the representatives of the logical database models. Based on these two basic database models, there exists a kind of hybrid database model called object-relational database model. In addition, new developments in artificial intelligence and procedure control have resulted in the appearances of deductive databases, active databases, temporal databases, and spatial databases. These databases generally adopt either one of the above-mentioned two basic database models or a hybrid database model. In this subchapter, several logical database models including relational database model, nested relational database model, and object-oriented database model will be introduced.

1.2.1 The Relational Database Model

Relational database model introduced first by Codd (1970) is the most successful one and relational databases have been extensively applied in most information systems in spite of the increasing populations of object-oriented databases. A relational database is a collection of relations.

1.2.1.1 Attributes and Domains

The representations for some features are usually extracted from real-world things. The features of a thing are called attributes. For each *attribute*, there exists a range that the attribute takes values, called *domain* of the attribute. A domain is a finite set of values and every value is an atomic data, the minimum data unit with meanings.

1.2.1.2 Relations and Tuples

Let $A_1, A_2, \ldots, A_n$ be attribute names and the corresponding attribute domains be $D_1, D_2, \ldots, D_n$ (or Dom(A_i), $1 \leq i \leq n$), respectively. Then relational schema R is represented as

$$R = (D_1/A_1, D_2/A_2, \ldots, D_n/A_n)$$

or

$$R = (A_1, A_2, \ldots, A_n),$$

where n is the number of attributes and is called the degree of relation.

The instances of R, expressed as r or r (R), are a set of n-tuples and can be represented as $r = \{t_1, t_2, \ldots, t_m\}$. A tuple t can be expressed as $t = \langle v_1, v_2, \ldots, v_n \rangle$, where $v \in D_i$ $(1 \leq i \leq n)$, i.e., $t \in D_1 \times D_2 \times \cdots \times D_n$. The quantity r is therefore a subset of Cartesian product of attribute domains, i.e., $r \subseteq D_1 \times D_2 \times \cdots \times D_n$. Viewed from the content of a relation, a relation is a simple table, where tuples are its rows and attributes are its columns. Note that there is no complex data in relational table. The value of a tuple t on attribute set S is generally written t [S], where $S \subseteq R$.

1.2.1.3 Keys

If an attribute value or the values of an attribute group in a relation can solely identify a tuple from other tuples, the attribute or attribute group is called a *super key* of the relation. If any proper subsets of a super key are not a super key, such super key is called a *candidate key* or shortly *key*.

For a relation, there may be several candidate keys. One chooses one candidate as the *primary key*, and other candidates are called *alternate key*. It is clear that the values of primary key of all tuples in a relation are different and are not null. The attributes included in a candidate key are called *prime attributes* and not included in any candidate key called *non-prime attributes*. If an attribute or an attribute group is not a key of relation *r* but it is a key of relation s, such attribute (group) is called *foreign key* of relation *r*.

1.2.1.4 Constraints

There are various constraints in the relational databases. We identify these constraints as follows.

(a) *Domain integrity constraints* The basic contents of domain integrity constraints are that attribute values should be the values in the domains. In addition, domain integrity constraints are also prescribed if an attribute value could be null.
(b) *Entity integrity constraints* Every relation should have a primary key and the value of the primary key in each tuple should be sole and cannot be null.
(c) *Referential integrity constraints* Let a relation *r* have a foreign key *FK* and the foreign key value of a tuple *t* in *r* be $t\,[FK]$. Let *FK* quote the primary key *PK* of relation r' and t' be a tuple in r'. Referential integrity constraint demands that $t\,[FK]$ comply with the following constraint: $t\,[FK] = t'\,[PK]$/null.
(d) *General integrity constraints* In addition to the above-mentioned three kinds of integrity constraints that are most fundamental in relational database model, there are other integrity constraints related to data contents directly, called *general integrity constraints*. Because numbers of them are very large, only a few of them are considered in current relational DBMSs. Among these constraints, *functional dependencies* (*FD*) and *multivalued dependencies* (*MVD*) are more important in relational database design theory and widely investigated.

The functional dependencies (*FD*) in relational databases represent the dependency relationships among attribute values in relation. In the relational databases, functional dependencies can be described as follows.

For a relation *r* (*R*), in which *R* denotes the schema, its attribute set is denoted by *U*, and $X, Y \subseteq U$, we say *r* satisfies the *functional dependency FD*: $X \rightarrow Y$, if $(\forall t \in r)\ (\forall s \in r)\ (t\,[X] = s\,[X] \Rightarrow t\,[Y] = s\,[Y])$.

Based on the concept of functional dependency, the partial/full functional dependencies and the transitive functional dependency can be described as follows.

For a relation *r* (*R*), in which *R* denotes the schema, its attribute set is denoted by U, and $X, Y \subseteq U$, we say *Y* is *fully functionally dependent* on *X*, denoted by $X \rightarrow_f Y$, if and only if $X \rightarrow Y$ and there does not exit $X' \subset X$ ($X' \neq \Phi$) such that $X' \rightarrow Y$. If such X' exits, then *Y* is *partially functionally dependent* on *X*, denoted by $X \rightarrow_p Y$.

The notion of keys can consequently be described in terms of *FDs*.

For a relation *r* (*R*), in which *R* denotes the schema, its attribute set is denoted by *U*, and $K \subseteq U$, we say *K* is a candidate key of *R* if and only if $K \rightarrow_f U$.

Multivalued dependencies (*MVD*) originated by Fagin (1977) are another important data dependencies that are imposed on the tuples of relational databases, relating an attribute value or a set of attribute values to a set of attribute values, independent of the other attributes in the relation. In classical relational databases, multivalued dependencies can be described as follows.

For a relation *r* (*R*), in which *R* denotes the schema, its attribute set is denoted by $U, X, Y \subseteq U$, and $Z = U - XY$, we say *r* satisfies the multivalued dependency *MVD*: $X \rightarrow \rightarrow Y$, if $(\forall\, t \in r)\, (\forall\, s \in r)\, (t\, [X] = s\, [X] \Rightarrow (\exists\, u \in r)\, (u\, [X] = t\, [X] \wedge u\, [Y] = t\, [Y] \wedge u\, [Z] = s\, [Z]))$.

In the relational databases, the functional and multivalued dependencies satisfy the inference rules, namely, the *axiom systems* (Armstrong 1974; Beeri et al. 1977). For the functional dependency, for example, the *Armstrong axioms* (1974) can be used to derive all possible *FDs* implied by a given set of dependencies. Let *r* (*R*) be a relation on schema *R*, its attribute set be denoted by *U* and *X*, *Y*, $Z \subseteq U$. Then the following is a set of Armstrong axioms.

(a) *Inclusion rule*: if $X \supseteq Y$, then $X \rightarrow Y$.
(b) *Transitivity rule*: if $X \rightarrow Y$ and $Y \rightarrow Z$, then $X \rightarrow Z$.
(c) *Augmentation rule*: If $X \rightarrow Y$, then $X \cup Z \rightarrow Y \cup Z$.

1.2.1.5 The Relational Algebra

Relational database model provides some operations, called the *relational algebra operations*. These operations can be subdivided into two classes:

(a) the operations for relations only (*select*, *project*, *join*, and *division*) and
(b) the set operations (*union*, *difference*, *intersection*, and *Cartesian product*).

In addition, some new operations such as *outerjoin*, *outerunion* and *aggregate* operations are developed for database integration or statistics and decision support. By using these operations, one can query or update relations.

Union (∪)

Union is a binary set operation on two relations that are union-compatible. That means they have the same number of attributes with the same domains pairwisely. Formally let *r* and *s* be two union-compatible relations on the scheme $R(A_1, A_2, \ldots, A_n)$. Then

$$r \cup s = \{t | t \in r \vee t \in s\}$$

It is clear that the result of $r \cup s$ is a relation on the schema R that includes all tuples which are either in r or in s or in both r and s. Of course, duplicate tuples, if any, must be eliminated.

Difference (−)

Difference is a binary set operation on two relations that are union-compatible. Formally let r and s be two union-compatible relations on the scheme $R(A_1, A_2, \ldots, A_n)$. Then

$$r - s = \{t | t \in r \wedge t \notin s\}$$

It can be seen that the result of $r - s$ is a relation on the schema R that includes the tuples which are only in r but not in s.

Cartesian product (×)

Cartesian product is a binary set operation on two relations. Formally let r and s be two fuzzy relations on schema R and S, respectively. Then

$$r \times s = \{t(R \cup S) \, | t[R] \in r \wedge t[S] \in s\}$$

That is, the result of $r \times s$ is a relation on the schema $R \cup S$, in which a tuple is a combination of a tuple from r and a tuple from s. So $|r \times s| = |r| \times |s|$. Here $|r|$ denotes the number of tuples in r.

Projection (Π)

Projection is a unary operation on a relation. Formally let r be relation on the scheme $R(A_1, A_2, \ldots, A_n)$. Then the projection of r over attribute subset S ($S \subset R$) is defined as follows.

$$\Pi_S(r) = \{t\,(S) | (\forall x)\,(x \in r\,(S) \wedge t = x[S])\}.$$

In other words, the result of Π_S (r) is a relation on the schema S that only includes the columns of relational table r which are given in S. It should noticed that, if the attributes in S are all non-key attributes of r (R), duplicate tuples may appear in Π_S (r) and must be eliminated.

Selection (σ)

Selection is a unary operation on a relation. Formally let r be relation on the scheme $R(A_1, A_2, \ldots, A_n)$. Then the selection of r based on a selection condition P specified by a Boolean expression in forms of a single or composite predicate is defined as follows.

$$\sigma_P\,(r) = \{t \,|\, t \in r \wedge P\,(t)\}.$$

Clearly, the result of σ_P (r) is a relation on the schema R that only includes the tuples in r which satisfy the given selection condition P.

The five relational operations given above are called the primitive operations in the relational databases. In addition, there are three additional relational operations, namely, intersection, join (and natural join), and division. But the three operations can be defined by the primitive operations.

Intersection ($\cap$)

Intersection is a binary set operation on two relations that are union-compatible. Formally let r and s be two union-compatible relations on the scheme $R(A_1, A_2, \ldots, A_n)$. Then

$$r \cap s = \{t | t \in r \wedge t \in s\} \text{ or } r \cap s = r - (r - s)$$

The result of $r \cap s$ is a relation on the schema R that includes the tuples which are both in r and in s.

Join ($\bowtie$)

Join is a binary set operation on two relations. Formally let r (R) and s (S) be any two relations. Let P be a conditional predicate in the form of $A\ \theta\ B$, where $\theta \in \{>, <, \geq, \leq, =, \neq\}$, where $A \in R$, and $B \in S$. Then

$$r \bowtie_P s = \{t\ (R \cup S) \mid t\ [R] \in r \wedge t\ [S] \in s \wedge P\ (t\ [R], t\ [S])\} \text{ or}$$
$$r \bowtie_P s = \sigma_P\ (r \times s)$$

The result of $r \bowtie_P s$ is a relation on the schema $R \cup S$, in which a tuple is a combination of a related tuple from r and a related tuple from s. Not being the same as the Cartesian product operation, the two combined tuples respectively from r and s must satisfy the given condition.

When attributes A and B are identical and "θ" takes $=$, the join operation becomes the natural join operation, denoted $r \bowtie s$. Let $Q = R \cap S$. Then

$$r \bowtie s = \{t(R \cup (S - Q)) | (\exists x)\ (\exists y)\ (x(R) \wedge y \in s(S) \wedge x[Q] = y[Q] \wedge t[R] = x[R] \wedge t[S - Q] = y[S - Q]\}$$

Division ($\div$)

Division, referred to quotient operation sometimes, is used to find out the sub-relation $r \div s$ of a relation r, containing sub-tuples of r which have for complements in r all the tuples of a relation s. Then the division operation is defined by

$$r \div s = \{t | (\forall u)(u \in s \wedge (t, u) \in r)\},$$

where u is a tuple of s and t is a sub-tuple of r such that (t, u) is a tuple of r.

Alternatively, let r (R) and s (S) be two relations, where $S \subset R$. Let $Q = R - S$. Then the division of r and s can be defined as follows.

$$r \div s = \Pi_Q(r) - \Pi_Q(r\,(Q) \times s - r)$$

1.2.1.6 Relational Database Design

1. *Overall Design of Databases*

The objective of database design is to capture the essential aspects of some real-world enterprise for which one wishes to construct a database (Petry 1996). Figure 1.1 shows a simplified description of the database design process (Elmasri

Fig. 1.1 Database design process

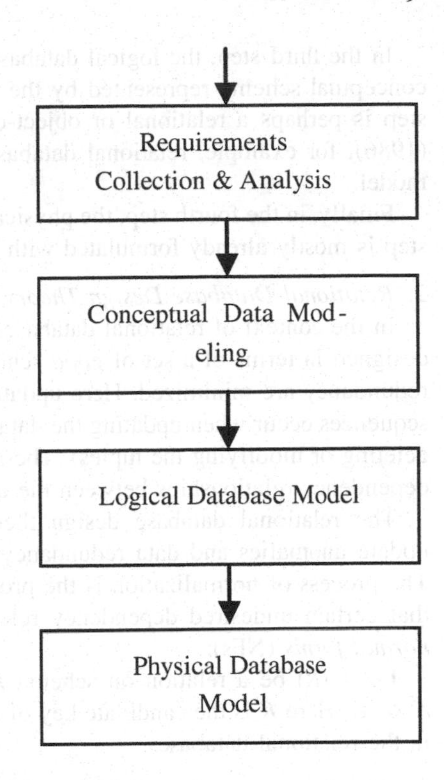

and Navathe 1994). Then four major steps are applied for the database design process, which are the *requirements collection and analysis*, *conceptual data modeling*, *logical database model*, and *physical database model*, respectively.

In the first step, the database designers collect and analyze the data requirements from prospective database users. As a result of this step, a concisely written set of users' requirements is formed.

In the second step, the conceptual data models (e.g., ER/EER and UML) are used to create a conceptual schema for the database. Here, the conceptual schema is a concise descriptions of the data requirements of the users and includes detailed descriptions of the data types, relationships, constraints, and etc. But there are no any implementation details in the conceptual schema. So it should be easy to share the conceptual schema with non-technical users. It is worth mentioning that a complex database is generally designed cooperatively by a design group and each member of the group may have different background. So using multiple conceptual data models to create the conceptual schema can facilitate the database designers with different background to design their conceptual data schemas easily by using one of the conceptual data models they are familiar with. But finally all these conceptual schemas designed by different members should be converted into a union conceptual schema. There are already some efforts for converting different conceptual schemas (Cherfi et al. 2002).

In the third step, the logical database model is designed through mapping the conceptual schema represented by the conceptual data model. The result of this step is perhaps a relational or object-oriented database model. In Teorey et al. (1986), for example, relational databases were logically designed using the ER model.

Finally, in the fourth step, the physical database model is design. Of course, this step is mostly already formulated with a commercial DBMS.

2. *Relational Database Design Theory*

In the context of relational databases, the relational database model should be designed in terms of a set of *good* schemas such that update anomalies and data redundancy are minimized. Here update anomalies mean that the undesired consequences occur when updating the data in the relational databases (e.g., inserting, deleting or modifying the tuples). The reason is that there exist certain undesired dependency relationships between the attributes of a relation.

The relational database design theory has been developed for minimizing update anomalies and data redundancy, which core is the normalization theory. The process of normalization is the process of relation schema decomposition so that certain undesired dependency relationships are removed to lead to certain *normal forms* (NFs).

Let r (R) be a relation on schema R, U be the attribute set of R, and X, K, $A \subseteq U$. Here K is the candidate key of R. Then we have the following major NFs in the relational databases.

(a) *The first normal form* (1NF): R is in 1NF, denoted by $R \in$ 1NF, if and only if every attribute value in r (R) is atomic.
(b) *The second normal form* (2NF): R is in 2NF, denoted by $R \in$ 2NF, if and only if $R \in$ 1NF and for any non-prime attribute A, $K \rightarrow_f A$.
(c) *The third normal form* (3NF): R is in 3NF, denoted by $R \in$ 3NF, if and only if $R \in$ 1NF and for any $X \rightarrow A$ $(A \nsubseteq X)$, either X is a superkey of R or A is a set of prime attributes.
(d) *The Boyce-Codd normal form* (BCNF): R is in BCNF, denoted by $R \in$ BCNF, if and only if $R \in$ 1NF and for any $X \rightarrow A$ $(A \nsubseteq X)$, either X is a superkey of R.

A lower NF can be normalized into a higher NF through relation schema decomposition (via projection). Figure 1.2 shows the details (Chen 1999).

It should be noticed that the schema decomposition should satisfy the following properties:

(a) *lossless-join* It means that the relation reconstructed from the resultant relations of the decomposition will be the same as the original relation with respect to information contents.
(b) *dependency-preservation* It means that the functional dependencies in the original relation are preserved by the resultant relations of the decomposition.

The four NFs discussed above are based on the functional dependency. In addition, there are other kinds of normal forms such as the *fourth normal form*

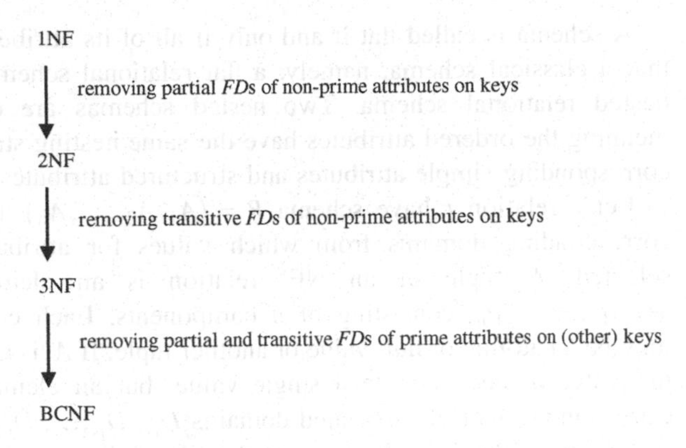

Fig. 1.2 Normal forms based on functional dependencies

(4NF) and the *fifth normal form* (5NF), which are related with multivalued dependency and join dependency, respectively.

1.2.2 The Nested Relational Database Model

The normalization, being one kind of constraints, is proposed in traditional relational databases. Among various normalized forms, first normal form (1NF) is the most fundamental one, which assumes that each attribute value in a relational instance must be atomic. As we know, the real-world applications are complex, and data types and their relationships are rich as well as complicated. The 1NF assumption limits the expressive power of traditional relational database model. Therefore, some attempts to relax 1NF limitation are made and one kind of data model, called non-first normal (or nested) relational database model have been introduced.

The first attempt to relax first-normal formal limitation is made by Makinouchi (1977), where, attribute values in the relation may be atomic or set-valued. Such relation is thereby called non-first normal form (NF^2) one. After Makinouchi's proposal, NF^2 database model is further extended (Ozsoyoglu et al. 1987; Schek and Scholl 1986). The NF^2 database model in common sense now means that attribute values in the relational instances are either atomic or set-valued and even relations themselves. So NF^2 databases are called nested relational databases also. Here, we do not differentiate these two notions. NF^2 relational schema can be described as follows.

An attribute A_j is a structured attribute if its schema appears on the left-hand side of a rule; otherwise it is simple. An NF^2 relational schema may contain any combination of simple or structured attributes on the right-hand side of the rules. Formally,

Schema:: Simple_attribute|Simple_attribute, Structured_attributes
Structured_attributes:: Simple_attribute|Simple_attribute, Structured_attributes

A schema is called flat if and only if all of its attributes are simple. It is clear that a classical schema, namely, a flat relational schema, is a special case of a nested relational schema. Two nested schemas are called union-compatible, meaning the ordered attributes have the same nesting structure, if and only if the corresponding simple attributes and structured attributes are union-compatible.

Let a relation r have schema $R = (A_1, A_2, \ldots, A_n)$ and let D_1, $D_2, \ldots$, D_n be corresponding domains from which values for attributes (A_1, $A_2, \ldots$, A_n) are selected. A tuple of an NF^2 relation is an element in r and denoted as $\langle a_1, a_2, \ldots, a_n \rangle$ consisting of n components. Each component a_j $(1 \leq j \leq n)$ may be an atomic or null value or another tuple. If A_j is a structured attribute, then the value a_j need not be a single value, but an element of the subset of the Cartesian product of associated domains D_{j1}, $D_{j2}, \ldots$, D_{jm}.

Based on the NF^2 database model, the ordinary relational algebra has been extended. In addition, two new restructuring operators, called the *Nest* and *Unnest* (Ozsoyoglu et al. 1987; Roth et al. 1987) [as well as *Pack* and *Unpack* (Ozsoyoglu et al. 1987)], have been introduced. The Nest operator can gain the nested relation including complex-valued attributes. The Unnest operator is used to flatten the nested relation. That is, it takes a relation nested on a set of attributes and desegregates it, creating a "flatter" structure. The formal definitions and the properties of these operators as well as the ordinary relational algebra for the NF^2 data model have been given (Colby 1990; Venkatramen and Sen 1993).

1.2.3 The Object-Oriented Database Model

Although there has been great success in using the relational databases for transaction processing, the relational databases have some limitations in some non-transaction applications such as computer-aided design and manufacturing (CAD/CAM), knowledge-based systems, multimedia, and GIS. Such limitations include the following.

(a) The data type is very restricted.
(b) The data structure based on the record notion may not match the real-world entity.
(c) Data semantics is not rich, and the relationships between two entities cannot be represented in a natural way.

Therefore, some non-traditional data models were developed in succession to enlarge the application area of databases since the end of the 1970s. Since these non-traditional data models appeared after the relational data model, they are called post-relational database models. Object-oriented database model is one of the post-relational database models.

Object-oriented (OO) data model is developed by adopting some concepts of semantic data models and knowledge expressing models, some ideas of object-oriented program language and abstract data type in data structure/programming.

1.2.3.1 Objects and Identifiers

All real-world entities can be simulated as *objects*, which have no unified and standard definition. Viewing from the structure, an object consists of *attributes*, *methods* and *constraints*. The attributes of an object can be simple data and other objects. The procedure that some objects constitute a new object is called *aggregation*. A method in an object contains two parts: signature of the method that illustrates the name of the method, parameter type, and result type; implementation of the method.

In general, attributes, methods and constraints in an object are encapsulated as one unit. The state of an object is changed only by passing message between objects. *Encapsulation* is one of the major features in OO data models.

In OO data models, each object has a sole and constant identifier, which is called *object identifier* (OID). For two objects with same attributes, methods and constraints, they are different objects if they have a different OID. The OID of an object is generated by system and cannot be changed by the user.

The OID generated by system can be divided into two kinds, i.e., *logical object identifier* and *physical object identifier*. Logical object identifier is mapped into physical one when an object is used because only physical object identifier concerns the storage address of the object.

1.2.3.2 Classes and Instances

In OO data models, objects with the same attributes, methods and constraints can be incorporated into a *class*, where objects are called *instances*. In a class, attributes, methods and constraints should be declared. Note that the attributes in a class can be classified into two kinds: instance variables and class variables. Instance variables are the attributes for which values are different in different objects of the class, while class variables are the attributes for which values are the same in different objects of the class.

In fact, classes can also be regarded as objects. Then, classes can be incorporated into another new class, called *meta class*. The instances of a meta class are classes. Therefore, objects are distinguished into *instance objects* and *class objects*.

1.2.3.3 Class Hierarchical Structure and Inheritance

A subset of a class, say *A*, can be defined as a class, say *B*. Class *B* is called a *subclass* and class *A* is called *superclass*. A subclass can further be divided into new subclasses. A *class hierarchical structure* is hereby formed, in which it is possible that a subclass may have multiple direct or indirect superclasses. The relationship between superclass and subclass is called *IS-A relationship*, which represents a *specialization* from top to bottom and a *generalization* from bottom to top. Because one subclass can have several direct superclasses, a class hierarchical structure is not a tree but a class *lattice*.

Because a subclass is a subset of its superclass, the subclass inherits the attributes and methods in its all superclasses. Besides inheritance, a subclass can define new attributes and methods or can modify the attributes and methods in the superclasses. If a subclass has several direct superclasses, the subclass inherits the attributes and methods from these direct superclasses. This is called *multiple inheritance*.

When inheriting, the naming conflict may occur, which should be resolved.

(a) Conflict among superclasses. If several direct superclasses of a subclass have the same name of attributes or methods, the conflict among superclasses appear. The solution is to declare the superclass order inherited, or to be illustrated by user.
(b) Conflict between a superclass and a subclass. When there are conflicts between a subclass and a superclass, the definition of attributes and methods in subclass would override the same definition in the superclass.

Note that a naming method may have a different meaning in different classes. The feature that a name has a multiple meaning is called *polymorphism*. The method with polymorphism is called *overloading*. Because the method in an object is polymorphism, the procedure corresponding to the method name cannot be determined while compiling program and do while running program. The later combination of the method name and implementing procedure of a method is called *late binding*.

1.3 Conceptual Data Models

Database systems are the key to implementing information modeling. Information modeling in databases can be carried out at two different levels: *conceptual data modeling* and *logical database modeling*. Correspondingly, we have *conceptual data models* and *logical database models* for information modeling. Generally the conceptual data models are used for information modeling at a high level of abstraction and at the level of data manipulation, i.e., a low level of abstraction, the logical database model is used for information modeling. Database modeling generally starts from the conceptual data models and then the developed conceptual data models are mapped into the logical database models.

The logical database models have been introduced in Sect. 1.2. In this section, we present several common conceptual data models. We briefly introduce the entity-relationship (ER) model, the enhanced (extended) entity-relationship (EER) model, and the class model of the Unified Modeling Language (UML).

1.3.1 Entity-Relationship and Enhanced ER Models

The entity-relationship (ER) model was incepted by Chen (1976) and has played a crucial role in database design and information systems analysis. In spite of its wide applications, the ER model suffers from its incapability of modeling complex objects and semantic relationships. So a number of new concepts have been introduced into the ER model by various researchers (dos Santos et al. 1979; Elmasri et al. 1985; Gegolla and Hohenstein 1991; Scheuermann et al. 1979) to enrich its usefulness and expressiveness, forming the notion of the enhanced entity-relationship (EER) model.

1.3.1.1 ER Model

The ER data model proposed by Chen (1976) can represent the real world semantics by using the notions of *entities*, *relationships*, and *attributes*. ER data schema described by the ER data model is generally represented by the *ER diagram*.

1. *Entity*

Entity is a concrete thing or an abstract notion that can be distinguishable and can be understood. A set of entities having the same characteristics is called an *entity set*. A named entity set can be viewed as the description of an *entity type*, while each entity in an entity set is an instance of that entity type. For example, "Car" is an entity set. The descriptions of the features of a car belong to the entity type, while an actual model car, for example, "Honda Civic DX", is an instance of the car entity. Sometimes entity type is called entity for short.

2. *Attribute and key*

The characteristics of an entity are called *attributes* of the entity. Each attribute has a range of values, called a *value set*. Value sets are essentially the same as attribute domains in relational databases.

Attributes in entities, however, can be not only simple attributes having one value set but also complex attributes having several value sets, called a *composite attribute*. For example, attributes "name" and "post address" of a person are a simple attribute and a complex attribute, respectively. In addition, an attribute can be *single-valued* or *multivalued*. For example, the attributes "Age" and "Email address" for a person are single-valued and multivalued attributes, respectively.

Like relational databases, a minimal set of attributes of an entity that can uniquely identify the entity is called a *key* of the entity. An entity may have more than one keys and one of them is designated as the *primary key*.

3. *Relationship*

Let $e_1, e_2, \ldots, e_n$ be entities. A relationship among them is represented as $r(e_1, e_2, \ldots, e_n)$. The relationship is 2-ary if $n = 2$ and is multiple-ary if $n > 2$. The set that consists of the same type of relationship is called *relationship set*. A relationship set can be viewed as a relationship among entity sets.

$R(E_1, E_2, ..., E_n)$ denotes the relationship set defined on entity sets $E_1, E_2, ..., E_n$. Relationship set is the type description of the entity relationship and a relationship among concrete entities is an instance of the corresponding relationship set. The same entity set can appear in a relationship set several times. A named relationship set can be viewed as the description of a *relationship type*. Sometimes relationship type is called relationship for short.

In the ER data model, a 2-ary relationship can be one-to-one, one-to-many, or many-to-many relationships. This classification can be applied to *n*-ary relationships as well. The constraint of a relationship among entities is called *cardinality ratio constraint*. In the ER data model there is an important semantic constraint called *participation constraint*, which stipulates the way that entities participate in the relationship. The concept *participation degree* is used to express the minimum number and maximum number of an entity participating in a relationship, expressed as (*min*, *max*) formally, where $max \geq min \geq 0$ and $max \geq 1$. When $min = 0$, the way an entity participates in a relationship is called *partial participation*, and is called *total participation* otherwise. The cardinality ratio constraint and participation constraint are, sometimes, referred to as the structure constraint.

Note that relationships in the ER data model also have attributes, called the *relationship attributes*.

There is a special relationship in the real world, which represents the ownership among entities and is called the *identifying relationship*. Such a relationship has two characteristics:

(a) The entity owned by another entity depends on an owning entity, and does not exist separately, which must totally participate in relationship.
(b) The entity owned by another entity may not be the entity key of itself.

Because the entity owned by another entity has such characteristics, it is called the *weak entity*. A weak entity can be regarded as an entity as well as a complex attribute of its owning entity.

4. *ER diagram*

In the ER diagram, entities, attributes and relationships should be represented, where a rectangle denotes an entity set, a rectangle with double lines denotes a weak entity set, and a diamond denotes a relationship. Rectangles and rhombus are linked by arcs and the cardinality ratios of relationships are given. If an arc is a single line, it represents that the entity is a partial participation. If an arc is a double line, it represents that the entity is a total participation. Participation degrees may be given if necessary.

In the ER diagram, a circle represents an attribute and it is linked to the corresponding entity set with an edge. If an attribute is an entity key or a part of the entity key, it is pointed out that in the ER diagram by underlining the attribute name or adding a short vertical line on the edge. If an attribute is complex, a tree structure will be formulated in the ER diagram.

Figure 1.3 shows ER diagram notations.

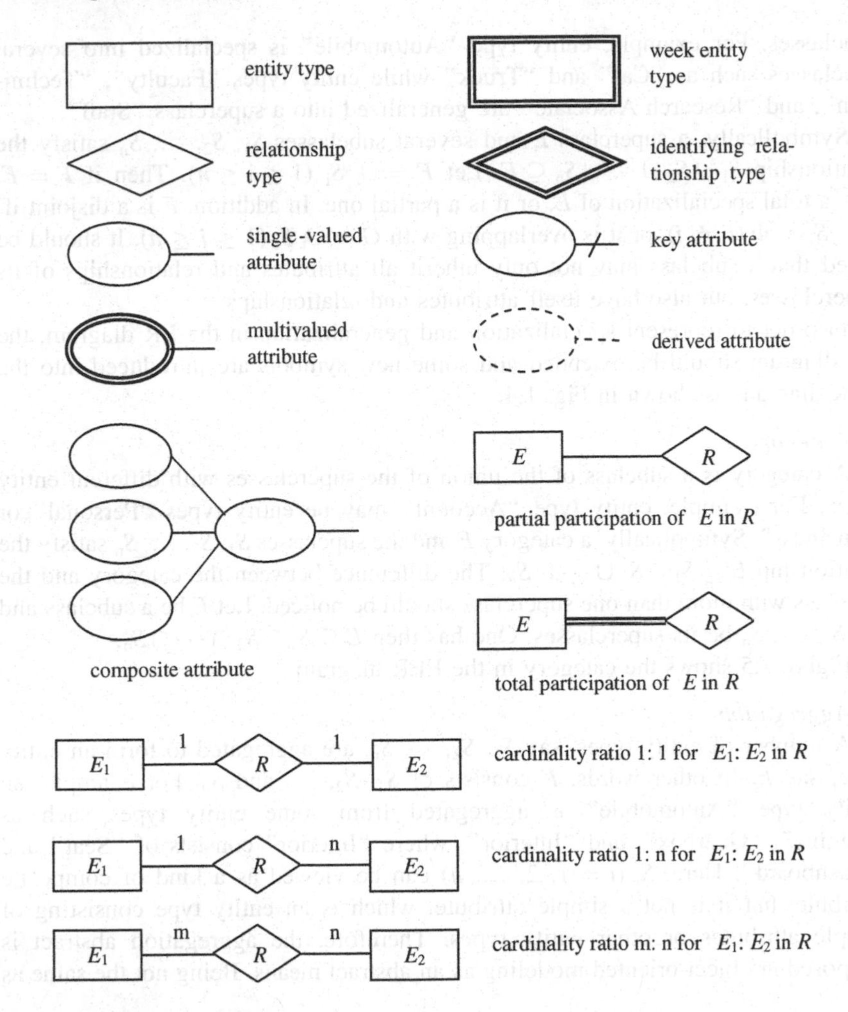

Fig. 1.3 ER diagram notations

1.3.1.2 EER Model

The ER model based on *entities*, *relationships* and *attributes* is called the basic ER model. In order to model the complex semantics and relationships in the applications such as CAD/CAM, CASE, GIS, and so on, some new concepts have been introduced in the ER model and the enhanced (extended) entity-relationship (EER) data model is formed. In the EER model, the following notions are introduced.

1. *Specialization and generalization*

Generalization can summarize several entity types with some common features to an entity type and define a superclass. Specialization can divide an entity type into several entity types according to a particular feature and define several subclasses. For example, entity type “Automobile” is specialized into several subclasses such as “Car” and “Truck” while entity types “Faculty”, “Technician”, and “Research Associate” are generalized into a superclass “Staff”.

Symbolically, a superclass E and several subclasses $S_1, S_2, \ldots, S_n$ satisfy the relationship $S_1 \cup S_2 \cup \cdots \cup S_n \subseteq E$. Let $F = \cup_i S_i$ ($1 \leq i \leq n$). Then if $F = E$, F is a total specialization of E, or it is a partial one. In addition, F is a disjoint if $S_i \cap S_j = \Phi$ ($i \neq j$), or it is overlapping with $G = \cup_i S_i$ ($1 \leq i \leq n$). It should be noted that a subclass may not only inherit all attributes and relationships of its superclasses, but also have itself attributes and relationships.

In order to represent specialization and generalization in the ER diagram, the ER diagram should be extended and some new symbols are introduced into the EER diagram as shown in Fig. 1.4.

2. *Category*

A category is a subclass of the union of the superclasses with different entity types. For example, entity type “Account” may be entity types “Personal” or “Business”. Symbolically, a category E and the supclasses $S_1, S_2, \ldots, S_n$ satisfy the relationship $E \subseteq S_1 \cup S_2 \cup \cdots \cup S_n$. The difference between the category and the subclass with more than one superclass should be noticed. Let E be a subclass and $S_1, S_2, \ldots, S_n$ be its superclasses. One has then $E \subseteq S_1 \cap S_2 \cap \cdots \cap S_n$.

Figure 1.5 shows the category in the EER diagram.

3. *Aggregation*

A number of entity types, say $S_1, S_2, \ldots, S_n$, are aggregated to form an entity type, say E. In other words, E consists of $S_1, S_2, \ldots,$ and S_n. For example, an entity type “Automobile” is aggregated from some entity types such as “Engine”, “Gearbox”, and “Interior”, where “Interior” consists of “Seat” and “Dashboard”. Here, S_i ($i = 1, 2, \ldots, n$) can be viewed as a kind of composite attribute, but it is not a simple attribute, which is an entity type consisting of simple attributes or other entity types. Therefore, the aggregation abstract is proposed in object-oriented modeling as an abstract means. Being not the same as

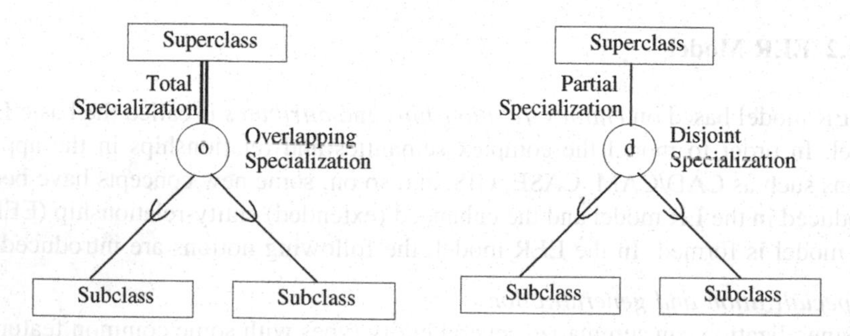

Fig. 1.4 EER diagram of the specialization

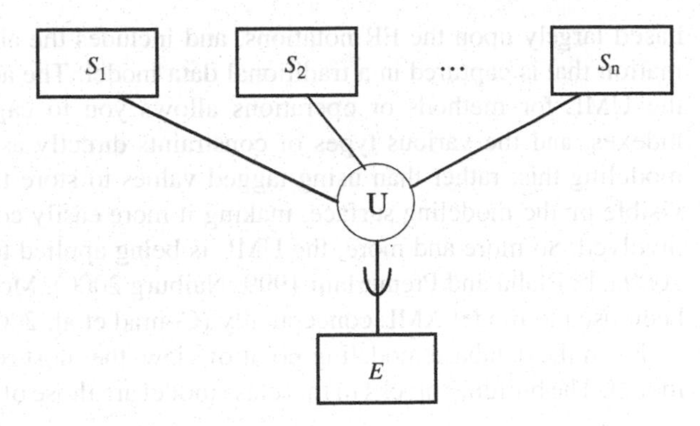

Fig. 1.5 EER diagram of the category

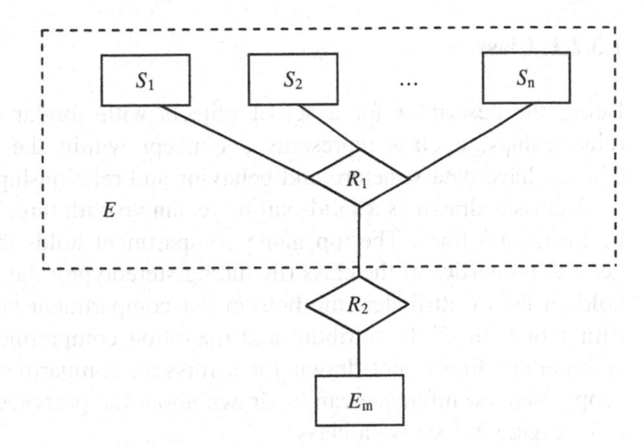

Fig. 1.6 EER diagram of the aggregation

specialization/generalization abstract, aggregated entity and all component entities belong to different entity types.

Figure 1.6 shows the aggregation in the EER diagram.

1.3.2 UML Class Model

The Unified Modeling Language (UML) (Booch et al. 1998; OMG 2001) is a set of OO modeling notations that has been standardized by the Object Management Group (OMG). The power of the UML can be applied for many areas of software engineering and knowledge engineering (Mili et al. 2001). The complete development of relational and object relational databases from business requirements can be described by the UML. The database itself traditionally has been described by notations called entity relationship (ER) diagrams, using graphic representation that is similar but not identical to that of the UML. Using the UML for database design has many advantages over the traditional ER notations (Naiburg 2000). The UML is based largely upon the ER notations, and includes the ability to capture all information that is captured in a traditional data model. The additional compartment in the UML for methods or operations allows you to capture items like triggers, indexes, and the various types of constraints directly as part of the diagram. By modeling this, rather than using tagged values to store the information, it is now visible on the modeling surface, making it more easily communicated to everyone involved. So more and more, the UML is being applied to data modeling (Ambler 2000a, b; Blaha and Premerlani 1999; Naiburg 2000). More recently, the UML has been used to model XML conceptually (Conrad et al. 2000).

From the database modeling point of view, the most relevant model is the class model. The building blocks in this class model are those of classes and relationships.

1.3.2.1 Class

Being the descriptor for a set of objects with similar structure, behavior, and relationships, a class represents a concept within the system being modeled. Classes have data structure and behavior and relationships to other elements.

A class is drawn as a solid-outline rectangle with three compartments separated by horizontal lines. The top name compartment holds the class name and other general properties of the class (including stereotype); the middle list compartment holds a list of attributes; the bottom list compartment holds a list of operations. Either or both of the attribute and operation compartments may be suppressed. A separator line is not drawn for a missing compartment. If a compartment is suppressed, no inference can be drawn about the presence or absence of elements in it. Figure 1.7 shows a class.

1.3.2.2 Relationships

Another main structural component in the class diagram of the UML is relationships for the representation of relationship between classes or class instances. UML supports a variety of relationships.

(a) *Aggregation and composition* An aggregation captures a whole-part relationship between an aggregate, a class that represent the whole, and a constituent part. An open diamond is used to denote an aggregate relationship. Here the class touched with the white diamond is the aggregate class, denoting the "whole". Figure 1.8 shows an aggregation relationship.

Fig. 1.7 The class icon of the UML

Class name
Attributes
Operations

Fig. 1.8 An aggregation relationship in the UML

Fig. 1.9 A generalization relation in the UML

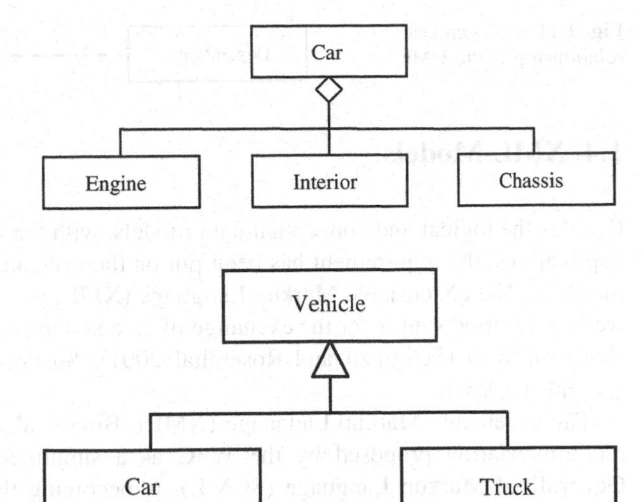

Aggregation is a special case of composition where constituent parts directly dependent on the whole part and they cannot exist independently. Composition mainly applies to attribute composition. A composition relationship is represented by a black diamond.

(b) *Generalization* Generalization is used to define a relationship between classes to build taxonomy of classes: one class is a more general description of a set of other classes. The generalization relationship is depicted by a triangular arrowhead. This arrowhead points to the superclass. One or more lines proceed from the superclass of the arrowhead connecting it to the subclasses. Figure 1.9 shows a generalization relationship.

(c) *Association* Associations are relationships that describe connections among class instances. An association is a more general relationship than aggregation or generalization. A role may be assigned to each class taking part in an association, making the association a directed link. An association relationship is expressed by a line with an arrowhead drawn between the participating classes. Figure 1.10 shows an association relationship.

(d) *Dependency* A dependency indicates a semantic relationship between two classes. It relates the classes themselves and does not require a set of instances for its meaning. It indicates a situation in which a change to the target class may require a change to the source class in the dependency. A dependency is shown as a dashed arrow between two classes. The class at the tail of the arrow depends on the class at the arrowhead. Figure 1.11 shows a dependency relationship.

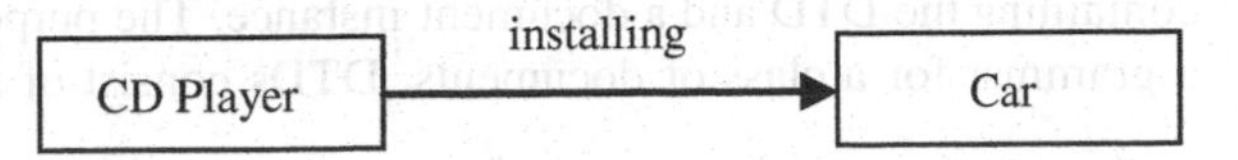

Fig. 1.10 An association relation in UML

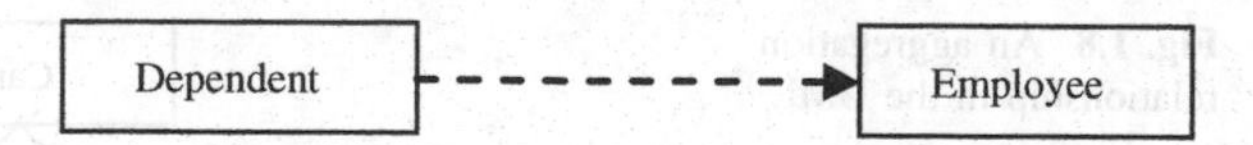

Fig. 1.11 A dependency relationship in the UML

1.4 XML Models

Besides the logical and conceptual data models, with the popularity of Web-based applications, the requirement has been put on the exchange and share of data over the Web. The eXtensiable Markup Language (XML) provides a Web-friendly and well-understood syntax for the exchange of data and impacts on data definition and share on Web (Seligman and Rosenthal 2001). So this subchapter will briefly present the XML.

The eXtensible Markup Language (XML) (Bray et al. 1998), a data formatting recommendation proposed by the W3C as a simplified form of the Standard Generalized Markup Language (SGML), is becoming the de facto standard for data description and exchange between various systems and databases over the Internet. As a new markup language, XML supports user-defined tags, encourages the separation of document content from its presentation, and is able to automate web information processing. This is creating a new set of data management requirements involving XML, such as the need to store and query XML documents.

1.4.1 XML Documents

An XML document has a logical and a physical structure (Bray et al. 1998). The physical structure is consists of entities that are ordered hierarchically. The logical structure is explicitly described by markups that comprise declarations, elements, comments, character references, and processing instructions.

XML documents that conform to the rules of XML mark-up are called "well-formed"; for example, each document must have a single top-level (root) element, and all tags must be correctly nested. A number of additional instructions arc permitted, such as comments, processing instructions, unparsed character data and entity references. Tags can also contain attributes in the form of name and values pairs, with the values enclosed in quotation marks. Figure 1.12 (Bourret 2004) shows an example XML document.

Essentially, XML documents can be associated with and validated against a schema specification in terms of a *document type definition* (*DTD*) (Bray et al. 1998) or by using the more powerful *XML* Schema language (Thompson et al. 2001; Biron and Malhotra 2001). In the following, we only focus on the DTD. Then XML document structure consists of an optional document type declaration containing the DTD and a document instance. The purpose of a DTD is to provide a grammar for a class of documents. DTDs consist of markup declarations.

```
<SalesOrder SONumber="12345">
   <Customer CustNumber="543">
         <CustName>ABC Industries</CustName>
         <Street>123 Main St.</Street>
         <City>Chicago</City>
         <State>IL</State>
         <PostCode>60609</PostCode>
   </Customer>
   <OrderDate>981215</OrderDate>
   <Item ItemNumber="1">
         <Part PartNumber="123">
               <Description>
                     <p><b>Turkey wrench:</b><br/>
                     Stainless steel, one-piece construction, lifetime guaran-
                     tee.</p>
               </Description>
               <Price>9.95</Price>
         </Part>
         <Quantity>10</Quantity>
   </Item>
   <Item ItemNumber="2">
         <Part PartNumber="456">
               <Description>
                     <p><b>Stuffing separator:<b><br/>
                     Aluminum, one-year guarantee.
                     </p>
               </Description>
               <Price>13.27</Price>
         </Part>
         <Quantity>5</Quantity>
   </Item>
</SalesOrder>
```

Fig. 1.12 Sales order XML document

1.4.2 XML DTD Constructs

According to the XML specification, DTDs consist of markup declarations namely element declarations, attribute-list declarations, entity declarations, notation declarations, processing instructions, and comments (Bray et al. 1998). As for these declarations, they are the elementary building blocks on which a DTD can be designed.

1. *Element Type and Attribute-list Declarations*
   Element type and attribute-list declarations make up the core of DTDs and declare the valid structures of a document instance, namely, the nested element tags with their additional attributes. An elements type declaration associates the element content. XML provides a variety of facilities for the construction of the element content, namely, *sequence* of elements, *choice* of elements, *cardinality* constructors (?, *, +), the types of EMPTY, ANY, #PCDATA, and *mixed* content. *Sequence* requires elements to have a fixed order, whereas *choice* expresses element alternatives. An EMPTY element has no content, whereas ANY indicates that the element can contain data of type #PCDATA or any other element defined in the DTD. Mixed is useful when elements are supposed to obtain character data (#PCDATA), optionally interspersed with child elements.
   The name of attribute list must match the name of the corresponding element. The list of attribute declaration consists of the attribute names, their types and default declarations.
2. *Entity Declarations*
   Entity declarations serve the reuse of DTD fragments and text as well as the integration of unparsed data. An entity declaration binds an entity to an identifier. Being external entities, unparsed entities always have notation references.
3. *Notation Declarations*
   Notation Declarations provide a name for the format of an unparsed entity. They might be used as reference in entity declarations, and in attribute-list declaration as well as in attribute specification.
4. *Processing Instructions*
   Processing instructions play an important role while checking integrity constraints of valid document instances. They have to be checked while parsing a document instance. The XML parse validates the document instance first and consumes the processing instructions known to XML. Then an application can handle more specific processing instructions.

A simple DTD of the XML document in Fig. 1.12 is given in Fig. 1.13 as follows.

It should be noted that, however, XML lacks sufficient power in modeling real-world data and their complex inter-relationships in semantics. Hence, it is necessary to use other methods to describe data paradigms and develop a true conceptual data model, and then transform this model into an XML encoded format, which can be treated as a logical model (Lee et al., 2001). Figure 1.14 depicts such a procedure to integrate conceptual data models and XML, making it easier to create, manage and retrieve XML documents.

Conceptual data modeling of XML schema [here XML schema refers to XML DTD or XML Schema, while XML Schema refers to the XML schema language proposed by W3C (Thompson et al. 2001; Biron and Malhotra 2001)] has been studied in the recent past. In Conrad et al. (2000), UML was used for designing

```
<!ELEMENT SaleOrder (Customer*, OrderDate, Item*)>
    <!ATTLIST SaleOrder SONumber IDREF #REQUIRED>
<!ELEMENT Customer (CustName?, Street?, City?, State?, PostCode?)>
    <!ATTLIST Customer CustNumber IDREF #REQUIRED>
<!ELEMENT CustName (#PCDATA)>
<!ELEMENT Street (#PCDATA)>
<!ELEMENT City (#PCDATA)>
<!ELEMENT State (#PCDATA)>
<!ELEMENT PostCode (#PCDATA)>
<!ELEMENT OrderDate (#PCDATA)>
<!ELEMENT Item (Part*, Quantity)>
    <!ATTLIST Item ItemNumber IDREF #REQUIRED>
<!ELEMENT Part (Description?, Price?)>
    <!ATTLIST Part PartNumber IDREF #REQUIRED>
<!ELEMENT Description (#PCDATA)>
<!ELEMENT Price (#PCDATA)>
<!ELEMENT Quantity (#PCDATA)>
```

Fig. 1.13 The DTD of the XML document in Fig. 1.12

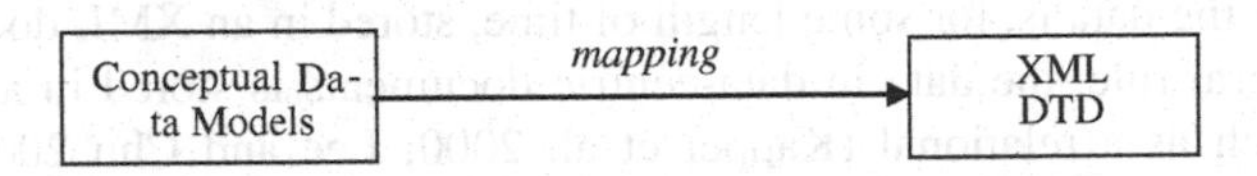

Fig. 1.14 Transformation from conceptual data models to XML DTD

XML DTD. The idea is to use essential parts of static UML to model XML DTD. The mapping between the static part of UML (i.e., class diagrams) and XML DTDs was developed. To take advantage of all facets that DTD concepts offer, the authors further extended the UML language in an UML-compliant way. Focusing on conceptual modeling at XML Schema level instead of XML DTD level, Xiao et al. (2001) introduced a solution for modeling XML and the transformation from object-oriented (OO) conceptual models to XML Schema, where the OO features are more general and are not limited to UML. Also in Mani et al. (2001), a set of features found in various XML schema languages (e.g., XML DTD and XML Schema) was formalized into XGrammar and the conversion between an XGrammar and EER model was presented. The EER was also used in Elmasri et al. (2002) to generate customized hierarchical views and then further create XML schemas from the hierarchical views. In Psaila (2000), the ER model was extended to ERX so that one can represent a style sheet and a collection of documents conforming to one DTD in ERX model. But order was represented in ERX model by an additional order attribute.

XML DTDs can also be converted to conceptual models. In dos Santos Mello and Heuser (2001), a semi-automatic process for converting an XML DTD to a schema in a canonical conceptual model based on ORM/NIAM and extended ER models was described. A set of conversion rules, which was the core of this process, was hereby developed.

1.4.3 XML Databases

It is crucial for Web-based applications to model, storage, manipulate, and manage XML data documents. XML documents can be classified into *data-centric documents* and *document-centric documents* (Bourret 2004).

1. *Data-Centric Documents*

Data-centric documents are characterized by fairly regular structure, fine-grained data (i.e., the smallest independent unit of data is at the level of a PCDATA-only element or an attribute), and little or no mixed content. The order in which sibling elements and PCDATA occurs is generally not significant, except when validating the document. Data-centric documents are documents that use XML as a data transport. They are designed for machine consumption and the fact that XML is used at all is usually superfluous. That is, it is not important to the application or the database that the data is, for some length of time, stored in an XML document.

As a general rule, the data in data-centric documents is stored in a traditional database, such as a relational (Kappel et al. 2000; Lee and Chu 2000), object-oriented (Chung et al. 2001), object-relational (Surjanto et al. 2000), or hierarchical database. The data can also be transferred from a database to a XML document (Vittori et al. 2001; Shanmugasundaram et al. 2001; Carey et al. 2000).

For the transfers between XML documents and databases, the mapping relationships between their architectures as well as their data should be created (Lee and Chu 2000; Surjanto et al. 2000). Note that it is possible to discard some information such as the document and its physical structure when transferring data between them. It must be pointed out, however, that the data in data-centric documents such as semistructured data can also be stored in a native XML database, in which a document-centric document is usually stored.

The sales order XML document shown in Fig. 1.12 is data-centric.

2. *Document-Centric Documents*

Document-centric documents are characterized by less regular or irregular structure, larger grained data (that is, the smallest independent unit of data might be at the level of an element with mixed content or the entire document itself), and lots of mixed content. The order in which sibling elements and PCDATA occurs is almost always significant. Document-centric documents are usually documents that are designed for human consumption. As a general rule, the documents in document-centric documents are stored in a native XML database or a content

```
<Product>
<Intro> The <ProductName>Turkey Wrench</ProductName> from <Developer>Full
Fabrication Labs, Inc.</Developer> is <Summary>like a monkey wrench,
but not as big.</Summary> </Intro>
<Description>
<Para>The turkey wrench, which comes in <i>both right- and left-
handed versions (skyhook optional)</i>, is made of the <b>finest
stainless steel</b>. The Readi-grip rubberized handle quickly adapts
to your hands, even in the greasiest situations. Adjustment is
possible through a variety of custom dials.</Para>
<Para>You can:</Para>
<List>
<Item><Link URL="Order.html">Order your own turkey wrench</Link></Item>
<Item><Link URL="Wrenches.htm">Read more about wrenches</Link></Item>
<Item><Link URL="Catalog.zip">Download the catalog</Link></Item>
</List>
<Para>The turkey wrench costs <b>just $19.99</b> and, if you
order now, comes with a <b>hand-crafted shrimp hammer</b> as a bonus gift.</Para>
</Description>
</Product>
```

Fig. 1.15 The document-centric XML document of product description

management system (an application designed to manage documents and built on top of a native XML database). Native XML databases are databases designed especially for storing XML documents. The only difference of native XML databases from other databases is that their internal model is based on XML and not something else, such as the relational model.

In practice, however, the distinction between data-centric and document-centric documents is not always clear. So the above-mentioned rules are not of a certainty. Data, especially semistructured data, can be stored in native XML databases and documents can be stored in traditional databases when few XML-specific features are needed. Furthermore, the boundaries between traditional databases and native XML databases are beginning to blur, as traditional databases add native XML capabilities and native XML databases support the storage of document fragments in external databases. The following product description given in Fig. 1.15 is document-centric (Bourret 2004).

1.5 Summary

How to manage a large number of data in various real-world application domains is particularly important. Databases and eXtensible Markup Language (XML) play essential roles for realize data management and information modeling. Various database models have been developed to operate and manage large quantities of data. Also, the XML is emerging and gradually considered as the de facto standard for data description and exchange between various systems and databases over the Internet. In this chapter, we introduce several common database models, including conceptual data models (entity-relationship (ER) model, enhanced entity-relationship (EER) model, and UML data model) and logical database models (relational database model and object-oriented database model). Also we discuss the data management requirements involving XML.

However, the traditional database models and XML feature limitations, mainly with what can be said about fuzzy information that is commonly found in many application domains. In order to provide the necessary means to handle and manage such information there are today a huge number of proposals for fuzzy extensions to database models and XML. In particular, Zadeh's *fuzzy set theory* (Zadeh 1965) has been identified as a successful technique for modelling the fuzzy information in many application areas, especially in the databases and XML. In Chap. 2, we will briefly introduce the fuzzy set theory and fuzzy database models.

References

Abiteboul S, Hull R (1987) IFO: a formal semantic database model. ACM Trans Database Syst 12(4):525–565

Abiteboul S, Hull R, Vianu V (1995) Foundations of databases. Addison Wesley, USA

Armstrong WW (1974) Dependency structures of data base relationships. In: Proceedings of the IFIP congress, pp 580–583

Ambler SW (2000a) The design of a robust persistence layer for relational databases. http://www.ambysoft.com/persistenceLayer.pdf

Ambler SW (2000b) Mapping objects to relational databases. http://www.AmbySoft.com/mappingObjects.pdf

Beeri C, Fagin R, Howard JH (1977) A complete axiomatization for functional and multivalued dependencies in database relations. In: Proceedings of the ACM SIGMOD conference, pp 47–61

Biron PV, Malhotra A (2001) XML schema Part 2: datatypes, W3C recommendation. http://www.w3.org/TR/xmlschema-2/

Blaha M, Premerlani W (1999) Using UML to design database applications. http://www.therationaledge.com/rosearchitect/mag/archives/9904/f8.html

Booch G, Rumbaugh J, Jacobson I (1998) The unified modeling language user guide. Addison-Welsley Longman Inc., USA

Bourret R (2004) XML and databases. http://www.rpbourret.com/xml/XMLAndDatabases.htm

Bray T, Paoli J, Sperberg-McQueen CM (1998) Extensible markup language (XML) 1.0, W3C recommendation. http://www.w3.org/TR/1998/REC-xml-19980210

Carey MJ, Kiernan J, Shanmugasundaram J, Shekita EJ, Subramanian SN (2000) XPERANTO: middleware for publishing object-relational data as XML documents. In: Proceedings of 26th international conference on very large data bases, pp 646–648

Chen PP (1976) The entity-relationship model: toward a unified view of data. ACM Trans Database Syst 1(1):9–36

Chen GQ (1999) Fuzzy logic in data modeling; semantics, constraints, and database design. Kluwer Academic Publisher, USA

Cherfi SS, Akoka J, Comyn-Wattiau I (2002) Conceptual modeling quality: from EER to UML schemas evaluation. Lect Notes Comput Sci 2503:414–428

Chung TS, Park S, Han SY, Kim HJ (2001) Extracting object-oriented database schemas from XML DTDs using inheritance. Lect Notes Comput Sci 2115:49–59

Codd EF (1970) A relational model of data for large shared data banks. Commun ACM 13(6):377–387

Colby LS (1990) A recursive algebra for nested relations. Inf Syst 15(5):567–662

Conrad R, Scheffiner D, Freytag JC (2000) XML conceptual modeling using UML. Lect Notes Comput Sci 1920:558–571

dos Santos Mello R, Heuser CA (2001) A rule-based conversion of a DTD to a conceptual schema. Lect Notes Comput Sci 2224:133–148

dos Santos C, Neuhold E, Furtado, A (1979) A data type approach to the entity-relationship model. In: Proceedings of the 1st international conference on the entity-relationship approach to systems analysis and design, pp 103–119

Elmasri R, Navathe SB (1994) Fundamentals of database systems. Second Edition, Benjamin/Cummings

Elmasri R, Weeldreyer J, Hevner A (1985) The category concept: an extension to the entity-relationship model. Int J Data Knowl Eng 1(1):75–116

Elmasri R, Wu YC, Hojabri B, Li C, Fu J (2002) Conceptual modeling for customized XML schemas. Lect Notes Comput Sci 2503:429–443

Fagin R (1977) Multivalued dependencies and a new normal form for relational databases. ACM Trans Database Syst 2(3):262–278

Gegolla M, Hohenstein U (1991) Towards a semantic view of an extended entity-relationship model. ACM Trans Database Syst 16(3):369–416

Hammer M, McLeod D (1981) Database description with SDM: a semantic database model. ACM Trans Database Syst 6(3):351–386

Kim W, Lochovsky FH (1989) Object-oriented concepts, databases and applications. Addison Wesley, USA

Kappel G, Kapsammer E, Rausch-Schott S, Retschitzegger W (2000) X-Ray: towards integrating XML and relational database systems. Lect Notes Comput Sci 1920:339–353

Lee DW, Chu WW (2000) Constraints-preserving transformation from XML document type definition to relational schema. Lect Notes Comput Sci 1920:323–338

Lee ML, Lee SY, Ling TW, Dobbie G, Kalinichenko LA (2001) Designing semistructured databases: a conceptual approach. Lect Notes Comput Sci 2113:12–21

Makinouchi A (1977) A consideration on normal form of not-necessarily normalized relations in the relational data model. In: Proceedings of the third international conference on very large databases. Tokyo, Japan, pp 447–453

Mani M, Lee DW, Muntz RR (2001) Semantic data modeling using XML schemas. Lect Notes Comput Sci 2224:149–163

Mili F, Shen W, Martinez I, Noel Ph, Ram M, Zouras E (2001) Knowledge modeling for design decisions. Artif Intell Eng 15:153–164

Naiburg E (2000) Database modeling and design using rational rose 2000. http://www.therationaledge.com/rosearchitect/mag/current/spring00/f5.html

Ozsoyoglu G, Ozsoyoglu ZM, Matos V (1987) Extending relational algebra and relational calculus with set-valued attributes and aggregate functions. ACM Trans Database Syst 12(4):566–592

OMG (2001) Unified modeling language (UML), version 1.4. http://www.omg.org/technology/documents/formal/uml.htm

Petry FE (1996) Fuzzy databases: principles and applications. Kluwer Academic Publisher, USA

Psaila G (2000) ERX: a data model for collections of XML Documents. In: Proceedings of the 2000 ACM symposium on applied computing, vol 2, pp 898–903

Roth MA, Korth HF, Batory DS (1987) SQL/NF: a query language for non-1NF relational databases. Inf Syst 12:99–114

Schek HJ, Scholl MH (1986) The relational model with relational-valued attributes. Inf Syst 11(2):137–147

Scheuermann P, Schiffner G, Weber H (1979) Abstraction capabilities and invariant properties modeling within to the entity-relationship approach. In: Proceedings of the 1st international conference on the entity-relationship approach to systems analysis and design, pp 121–140

Seligman L, Rosenthal A (2001) XML's impact on databases and data sharing. IEEE Comput 34(6):59–67

Shanmugasundaram J, Shekita EJ, Barr R, Carey MJ, Lindsay BG, Pirahesh H, Reinwald B (2001) Efficiently publishing relational data as XML documents. VLDB J 10(2–3):133–154

Surjanto B, Ritter N, Loeser H (2000) XML content management based on object-relational database technology. In: Proceedings of the first international conference on web information systems engineering, vol 1, pp 70–79

Teorey TJ, Yang DQ, Fry JP (1986) A logical design methodology for relational databases using the extended entity-relationship model. ACM Comput Surv 18(2):197–222

Thompson HS, Beech D, Maloney M, Mendelsohn N (2001) XML schema Part 1: structures, W3C recommendation. http://www.w3.org/TR/xmlschema-1/

Venkatramen SS, Sen A (1993) Formalization of an IS-A based extended nested relation data model. Inf Syst 20(1):53–57

Vittori CM, Dorneles CF, Heuser CA (2001) Creating XML documents from relational data sources. Lect Notes Comput Sci 2115:60–70

Xiao RG, Dillon TS, Chang E, Feng L (2001) Modeling and transformation of object-oriented conceptual models into XML schema. Lect Notes Comput Sci 2113:795–804

Zadeh LA (1965) Fuzzy sets. Inf Control 8(3):338–353

Chapter 2
Fuzzy Sets and Fuzzy Database Models

Abstract Information imprecision and uncertainty exist in many real-world applications, and for this reason fuzzy data modeling has been extensively investigated in various database models. In particular, Zadeh's fuzzy set theory has been identified as a successful technique for modeling imprecise and uncertain information in various database models. This has resulted in numerous contributions, mainly with respect to the popular fuzzy conceptual data models (fuzzy ER/EER model, fuzzy UML data model, and etc.) and fuzzy logical database models (fuzzy relational database model, fuzzy object-oriented database model, and etc.). Also, it is shown that fuzzy set theory is very useful in Web-based business intelligence. Therefore, topics related to the modeling of fuzzy data are considered very interesting in XML since it is the current standard data representation and exchange format over the Web. In particular, to manage fuzzy XML data, it is necessary to integrate fuzzy XML and various fuzzy databases, and various fuzzy database models (fuzzy relational database model and fuzzy object-oriented database model) need to be used for mapping to and from the fuzzy XML models. Therefore, in this chapter, we mainly introduce several fuzzy database models, including fuzzy UML data model, fuzzy relational database model, and fuzzy object-oriented database model. Before that, we briefly introduce some notions of fuzzy set theory.

2.1 Introduction

Information is often imprecise and uncertain in many real-world applications, and many sources can contribute to the imprecision and uncertainty of data or information. Therefore, it has been pointed out that we need to learn how to manage data that is imprecise or uncertain (Dalvi and Suciu 2007).

Unfortunately, the classical data management techniques such as databases and XML as introduced in Chap. 1 often suffer from their incapability of representing and manipulating imprecise and uncertain data information. On this basis, since

L. Yan et al., *Fuzzy XML Data Management*, Studies in Fuzziness and Soft Computing 311, DOI: 10.1007/978-3-642-44899-7_2,

the early 1980s, Zadeh's fuzzy logic (Zadeh 1965) has been introduced into various database models in order to enhance the classical models such that uncertain and imprecise information can be represented and manipulated. Over the past 30 years, a significant body of research in the area of fuzzy database modeling has been developed and tremendous gain is hereby accomplished in this area. Various fuzzy database models have been proposed, and some major issues related to these models have been investigated. In particular, fuzzy information has been extensively investigated in the context of the relational model (Petry 1996; Chen 1999). Recent efforts have extended these results to object-oriented databases by introducing the related notions of classes, generalization/specialization, and inheritance (de Caluwe 1998; Ma 2005a). In addition, fuzzy data modeling has been investigated in the context of the conceptual data models such as ER (Zvieli and Chen 1986), EER (Chen and Kerre 1998) and UML (Ma and Yan 2007; Ma et al. 2011; Haroonabadi and Teshnehlab 2007, 2009; Sicilia et al. 2002). More recently, XML data management is increasingly receiving attention due to the extensive use of Internet. Fuzzy information modeling in XML is hereby one of the foundations of implementing Web-based intelligent information processing (Ma 2005b).

In general, being similar to the classical database models, two kinds of database models can be identified, which are fuzzy conceptual data models (e.g., fuzzy ER (entity-relationship)/EER (enhanced/extended entity-relationship) models and fuzzy UML data model) and fuzzy logical database models (fuzzy relational database model and fuzzy object-oriented database model). Fuzzy conceptual data models for conceptual data modeling provide the designers with powerful mechanisms in generating the most complete specification from the real world. Fuzzy logical database models are often created through mapping fuzzy conceptual data models into fuzzy logical database models.

In this chapter, we mainly introduce several fuzzy database models, including fuzzy UML conceptual data model, fuzzy relational and fuzzy object-oriented logical database models. These models can be used for mapping to and from the fuzzy XML models in order to realize the fuzzy data management in many areas, such as database and Web-based application domains. Before that, we briefly introduce some notions of fuzzy set theory.

2.2 Imperfect Information and Fuzzy Sets

In real-world applications, information is often imperfect. Zadeh's fuzzy set theory has been identified as a successful technique for modeling imprecise and uncertain information in many application areas. Fuzzy set theory (Zadeh 1965), which is also interchangeably referred as fuzzy logic, is a generalization of the set theory and provides a means for the representation of imprecision and vagueness. One of the important areas of research in data management has been the continuous effort to enrich existing data management techniques (e.g., databases and XML) with a more extensive collection of semantic concepts. One of the semantic needs not adequately addressed by traditional models is that of imprecision and uncertainty. Traditional models assume the database model and XML to be a correct reflection of the world being captured and assume that the data stored is known, accurate, and complete. It is rarely the case in real life that all or most of these assumptions are met. Different fuzzy database models and fuzzy XML have been proposed to handle different categories of data quality (or lack thereof) with fuzzy set theory. Therefore, in this section, we briefly introduce some notions of fuzzy sets and possibility theory.

2.2.1 Imperfect Information

In order to satisfy the need for modeling fuzzy information, there have been some attempts at classifying various possible kinds of imperfect information. Inconsistency, imprecision, vagueness, uncertainty, and ambiguity are five basic kinds of imperfect information (Bosc and Prade 1993; Motor 1990; Motor and Smets 1997; Ma and Yan 2008; Smets 1997).

(a) Inconsistency is a kind of semantic conflict, meaning the same aspect of the real world is irreconcilably represented more than once in a database or in several different databases. For example, the *age* of *George* is stored as 34 and 37 simultaneously.
(b) Intuitively, imprecision and vagueness are relevant to the content of an attribute value, and it means that a choice must be made from a given range (interval or set) of values but it is not known exactly which one to choose per se. For example, *between 20 and 30 years old* and *young* for the attribute *Age* are imprecise and vague values, respectively. In general, vague information is represented by linguistic terms.
(c) Uncertainty is related to the degree of truth of its attribute value, meaning that we can apportion some, but not all, of our belief to a given value or group of values. For example, the possibility that the *age* of *Chris* is 35 right now should be 0.98. The random uncertainty, described using probability theory, is not considered here.
(d) Ambiguity means that some elements of the model lack complete semantics, leading to several possible interpretations.

In general, several different kinds of imperfect information can co-exist with respect to the same piece of information. In addition, imprecise values generally denote a set of values in the form of [ai1, ai2, …, aim] or [ai1, ai2] for the discrete and continuous universe of discourse, respectively, meaning that exactly one of the values is the true value for the single-valued attribute, or at least one of the values is the true value for the multivalued attribute. So, imprecise information here has two interpretations: disjunctive information and conjunctive information.

One kind of imprecise information that has been studied extensively is the well-known null values (Codd 1986, 1987; Motor 1990; Parsons 1996; Zaniolo 1984), which were originally called incomplete information. The possible interpretations of null values include: (a) *existing but unknown*, (b) *nonexisting* or *inapplicable*, and (c) *no information*. A null value on a multivalued object, however, means an "open null value" (Gottlob and Zicari 1988), i.e., the value may not exist, has exactly one unknown value, or has several unknown values. Null values with the semantics of "existent but unknown" can be considered as the special type of partial values that the true value can be any one value in the corresponding domain, i.e., an applicable null value corresponds to the whole domain.

The notion of a partial value is illustrated as follows (Grant 1979). A partial value on a universe of discourse U corresponds to a finite set of possible values in which exactly one of the values in the set is the true value, denoted by $\{a_1, a_2, \ldots, a_m\}$ for discrete U or $[a_1, a_n]$ for continua U, in which $\{a_1, a_2, \ldots, a_m\} \subseteq U$ or $[a_1, a_n] \subseteq U$. Let η be a partial value, then *sub* (η) and *sup* (η) are used to represent the minimum and maximum in the set.

Note that crisp data can also be viewed as special cases of partial values. A crisp data on discrete universe of discourse can be represented in the form of $\{p\}$, and a crisp data on continua universe of discourse can be represented in the form of $[p, p]$. Moreover, a partial value without containing any element is called an *empty partial value*, denoted by $\perp$. In fact, the symbol $\perp$ means an inapplicable missing data (Codd 1986, 1987). Null values, partial values, and crisp values are thus represented with a uniform format.

2.2.2 Fuzzy Sets

The fuzzy set was originally presented by Zadeh (1965). Since then fuzzy set has been infiltrating into almost all branches of pure and applied mathematics that are set-theory-based. This has resulted in a vast number of real applications crossing over a broad realm of domains and disciplines. Over the years, many of the existing approaches dealing with imprecision and uncertainty are based on the theory of fuzzy sets.

Let U be a universe of discourse. A fuzzy value on U is characterized by a fuzzy set F in U. A membership function

$$\mu_F : U \rightarrow [0, 1]$$

is defined for the fuzzy set F, where $\mu_F(u)$, for each $u \in U$, denotes the degree of membership of u in the fuzzy set F. For example, $\mu_F(u) = 0.8$ means that u is "likely" to be an element of F by a degree of 0.8. For ease of representation, a fuzzy set F over universe U is organized into a set of ordered pairs:

$$F = \{\mu_F(u_1)/u_1, \mu_F(u_2)/u_2, \ldots, \mu_F(u_n)/u_n\}.$$

When the membership function $\mu_F(u)$ above is explained to be a measure of the possibility that a variable X has the value u in this approach, where X takes values in U, a fuzzy value is described by a possibility distribution π_X (Zadeh 1978).

$$\pi_X = \{\pi_X(u_1)/u_1, \pi_X(u_2)/u_2, \ldots, \pi_X(u_n)/u_n\}$$

Here, $\pi_X(u_i)$, $u_i \in U$, denotes the possibility that u_i is true.

In addition, a fuzzy data is represented by similarity relations in domain elements (Buckles and Petry 1982), in which the fuzziness comes from the similarity relations between two values in a universe of discourse, not from the status of an object itself. Similarity relations are thus used to describe the degree similarity of two values from the same universe of discourse. A similarity relation *Sim* on the universe of discourse U is a mapping: $U \times U \rightarrow [0, 1]$ such that:

(i) for $\forall x \in U$, $Sim(x, x) = 1$, (reflexivity);
(ii) for $\forall x, y \in U$, $Sim(x, y) = Sim_i(y, x)$, (symmetry); and
(iii) for $\forall x, y, z \in U$, $Sim(x, z) \geq \max_y(\min(Sim(x, y), Sim(y, z)))$, (transitivity).

Moreover, the following notions related to fuzzy sets can be defined.

Support The set of the elements that have non-zero degrees of membership in F is called the support of F, denoted by

$$supp(F) = \{u \mid u \in U \text{ and } \mu_F(u) > 0\}.$$

Kernel The set of the elements that completely belong to F is called the kernel of F, denoted by

$$ker(F) = \{u \mid u \in U \text{ and } \mu_F(u) = 1\}.$$

α-Cut The set of the elements which degrees of membership in F are greater than (greater than or equal to) α, where $0 \leq \alpha < 1$ ($0 < \alpha \leq 1$), is called the strong (weak) α-cut of F, respectively denoted by

$$F_{\alpha+} = \{u \mid u \in U \text{ and } \mu_F(u) > \alpha\} \text{ and}$$
$$F_\alpha = \{u \mid u \in U \text{ and } \mu_F(u) \geq \alpha\}.$$

In addition, to manipulate fuzzy sets and possibility distributions, several common *set* operations are defined. The usual *set operations* (such as union, intersection and complementation) have been extended to deal with fuzzy sets (Zadeh 1965). Let A and B be fuzzy sets on the same universe of discourse U with the membership functions μ_A and μ_B, respectively. Then we have

Union The union of fuzzy sets A and B, denoted $A \cup B$, is a fuzzy set on U with the membership function $\mu_{A\cup B}$: $U \rightarrow [0, 1]$, where

$$\forall u \in U, \mu_{A\cup B}(u) = \max(\mu_A(u), \mu_B(u)).$$

Intersection The intersection of fuzzy sets A and B, denoted $A \cap B$, is a fuzzy set on U with the membership function $\mu_{A\cap B}$: $U \rightarrow [0, 1]$, where

$$\forall u \in U, \mu_{A\cap B}(u) = \min(\mu_A(u), \mu_B(u)).$$

Complementation The complementation of fuzzy set $\bar{A}$, denoted by $\bar{A}$, is a fuzzy set on U with the membership function $\mu_{\bar{A}}$: $U \rightarrow [0, 1]$, where

$$\forall u \in U, \mu_{\bar{A}}(u) = 1 - \mu_A(u).$$

Based on these definitions, the *difference* of the fuzzy sets B and A can be defined as:

$$B - A = B \cap \bar{A}.$$

Also, most of the properties that hold for classical set operations, such as DeMorgan's Laws, have been shown to hold for fuzzy sets. The only law of ordinary set theory that is no longer true is the law of the excluded middle, i.e.,

$$A \cap \bar{A} \neq \emptyset \text{ and } A \cup \bar{A} \neq U.$$

Let A, B and C be fuzzy sets in a universe of discourse U. Then the operations on fuzzy sets satisfy the following conditions:

- Commutativity laws: $A \cup B = B \cup A, A \cap B = B \cap A$
- Associativity laws: $(A \cup B) \cup C = A \cup (B \cup C)$, $(A \cap B) \cap C = A \cap (B \cap C)$
- Distribution laws: $(A \cup B) \cap C = (A \cap C) \cup (B \cap C)$, $(A \cap B) \cup C = (A \cup C) \cap (B \cup C)$
- Absorption laws: $(A \cup B) \cap A = A$, $(A \cap B) \cup A = A$
- Idempotency laws: $A \cup A = A, A \cap A = A$
- de Morgan laws: $\overline{A \cup B} = \bar{A} \cap \bar{B}$, $\overline{A \cap B} = \bar{A} \cup \bar{B}$

Given two fuzzy sets A and B in U, B is a fuzzy subset of A, denoted by $B \subseteq A$, if

$$\mu_B(u) \leq \mu_A(u) \text{ for all } u \in U.$$

Two fuzzy sets A and B are said to be equal if $A \subseteq B$ and $B \subseteq A$.

In order to define Cartesian product of fuzzy sets, let $U = U_1 \times U_2 \times \cdots \times U_n$ be the Cartesian product of n universes and $A_1, A_2, \ldots, A_n$ be fuzzy sets in $U_1, U_2, \ldots, U_n$, respectively. The Cartesian product $A_1 \times A_2 \times \cdots \times A_n$ is defined to be a fuzzy subset of $U_1 \times U_2 \times \cdots \times U_n$, where

$$\mu_{A1\times\cdots\times An}(u_1 \ldots u_n) = \min(\mu_{A1}(u_1), \ldots, \mu_{A\,n}(u_n))$$

and $u_i \in U_i, i = 1, \ldots, n$.

2.3 Fuzzy UML Data Models

Databases are the key to implementing data management. The conceptual data models play an important role in conceptual data modeling and database conceptual design. In general, the conceptual data models are used for information modeling at a high level of abstraction and at the level of data manipulation, i.e., a low level of abstraction. Database modeling generally starts from the conceptual data models and then the developed conceptual data models are mapped into the logical database models. Therefore, the conceptual data modeling may be the basic for data management. However, classical data models often suffer from their incapability to represent and manipulate imprecise and uncertain information that may occur in many applications in the real world.

In order to deal with complex objects and imprecise and uncertain information in conceptual data modeling, one needs fuzzy extension to conceptual data models, which allow imprecise and uncertain information to be represented and manipulated at a conceptual level. Various fuzzy conceptual data models have been proposed in the literature such as fuzzy ER model, fuzzy EER model, fuzzy IFO model, and fuzzy UML data model (Ma and Yan 2010). Note that, among these fuzzy conceptual data models, the fuzzy UML data model has the relatively strong expression. The fuzzy UML data model is a fuzzy extension of UML (Unified Modeling Language), which is a set of OO (object-oriented) modeling notations that has been standardized by the Object Management Group (OMG). In recent years, fuzzy UML data models were widely investigated for modeling imprecise and uncertain information (Ma and Yan 2007; Ma et al. 2011; Haroonabadi and Teshnehlab 2007, 2009; Sicilia et al. 2002). In this section, we focus on the fuzzy UML data model.

2.3.1 Fuzzy Class

An object, which is an entity of the real world, is fuzzy because of a lack of information. Formally, objects that have at least one attribute whose value is a fuzzy set are fuzzy objects. The objects with the same structure and behavior are grouped into a class, and classes are organized into hierarchies. Theoretically, a class can be considered from two viewpoints:

- An *extensional* class, where the class is defined by a list of its object instances;
- An *intensional* class, where the class is defined by a set of attributes and their admissible values.

In a fuzzy UML data model, a class is fuzzy because of the following several reasons (Ma and Yan 2007):

- A class is extensionally defined, where some objects with similar properties are fuzzy ones. The class defined by these objects may be fuzzy, and these objects belong to the class with degree of [0, 1].
- When a class is intensionally defined, the domains of some attributes may be fuzzy, and thus a fuzzy class is formed.
- The subclass produced by a fuzzy class by means of specialization, and the superclass produced by some classes (in which there is at least one class who is fuzzy) by means of generalization, are fuzzy.

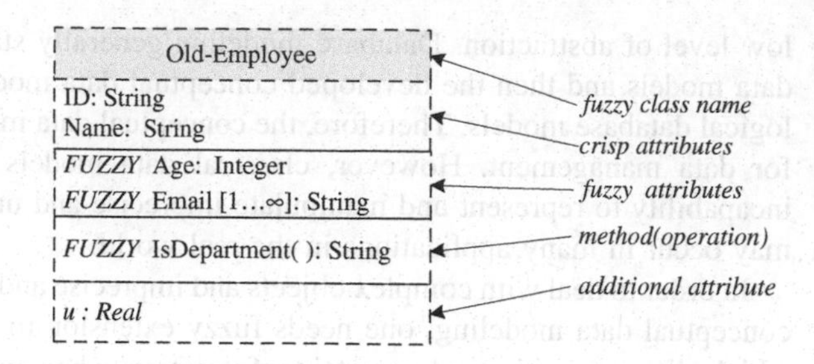

Fig. 2.1 Representation of a *fuzzy class Old-Employee* in a fuzzy UML data model

Figure 2.1 shows a fuzzy UML class *Old-Employee*, where a fuzzy class is denoted by using a *dashed-outline rectangle* to differentiate a classical class. Here:

- A fuzzy keyword *FUZZY* is appeared in front of an attribute indicating the attribute may take fuzzy values. For example, *FUZZY* Age or *FUZZY* Email.
- For an attribute a of type T in a class C, an optional multiplicity $[i\ldots j]$ for a specifies that a associates to each instance of C at least i and most j instances of T. When the multiplicity is missing, [1…1] is assumed, i.e., the attribute is single-valued. For example, the attribute "*FUZZY* Email [1…∞]" in Fig. 2.1 means that each object instance of the class *Old-Employee* has at least one email, and possibly more.
- The method *IsDepartment():String* denotes the dynamic aspect of fuzzy UML data models. It returns a possibility distribution value. The type of the parameter is *null*.
- An additional attribute $u \in [0, 1]$ in the class is defined for representing the object instance membership degree to the class.

2.3.2 Fuzzy Association

An *association* is a relation between the instances of two or more classes. Names of associations are unique in a fuzzy UML data model. An association has a related *association class* that describes properties of the association. Three kinds of fuzziness can be identified in an association relationship:

(i) The association is fuzzy itself, it means that the association relationship fuzzily exists in n classes, namely, this association relationship occurs with a degree of possibility.
(ii) The association is not fuzzy itself, i.e., it is known that the association relationship must occur in n classes. But it is not known for certain if n class instances (i.e., n objects) respectively belonging to the n classes have the given association relationship.
(iii) The association is fuzzy caused by such fact that (i) and (ii) occur in the association relationship simultaneously.

A Fuzzy Association relationship is an association relationship with a degree of possibility. We introduce a symbol β, as shown in Fig. 2.2, into a fuzzy UML data

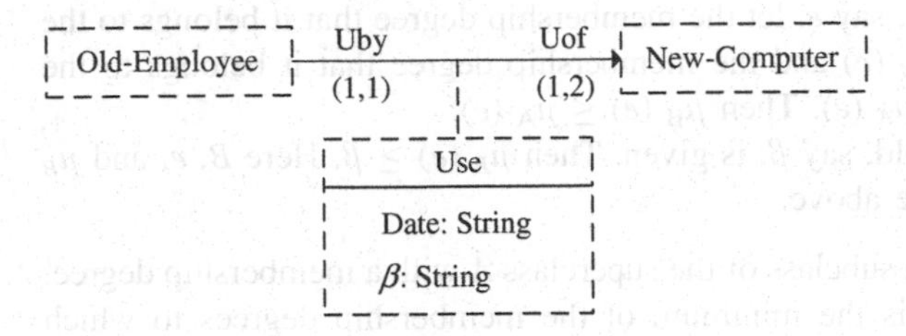

Fig. 2.2 Representation of a *fuzzy association* (*class*) in a fuzzy UML data model

model to denote the degree of possibility of a Fuzzy Association, and the calculating methods of β with respect to the three kinds of fuzziness above have been introduced in Ma et al. (2004).

Figure 2.2 shows a Fuzzy Association class *Use* between two classes *Old-Employee* and *New-Computer*. A *single line with an arrowhead* is used to denote a Fuzzy Association, and the association class is connected with the association by a *dashed-outline*. Here:

- *Date* is an attribute of the association class *Use*, which describes the start date that an *Old-Employee* uses a *New-Computer*.
- The additional symbol β denotes the membership degree of the Fuzzy Association occurring in several classes as mentioned above.
- The participation of a class in a Fuzzy Association is called a *role* which has a unique name. For example, *Uby* and *Uof* in Fig. 2.2.
- The cardinality constraint (*m*, *n*) on an association *S* specifies that each instance of the class C_i can participate at least *m* times and at most *n* times to *S*. For example, in Fig. 2.2, (1, 1) and (1, 2) denote that each *Old-Employee* can use at least 1 and at most 2 *New-Computers* and each *New-Computer* can be used by exactly one *Old-Employee*.

2.3.3 Fuzzy Generalization

The concept of *generalization* is one of the basic building blocks of the fuzzy UML data model. A *generalization* is a taxonomic relationship between a more general classifier named *superclass* and a more specific classifier named subclass. The subclass is produced from the superclass by means of inheriting all attributes and methods of the superclass, overriding some attributes and methods of the superclass, and defining some attributes and methods.

However, a class produced from a fuzzy class must be fuzzy. If the former is still called subclass and the later superclass, the subclass/superclass relationship is fuzzy. In other words, a class is a subclass of another class with membership degree of [0, 1] at this moment. We have the following criteria to determine the fuzzy subclass/superclass relationship.

(a) For any (fuzzy) object, say e, let the membership degree that it belongs to the subclass, say B, be $\mu_B(e)$ and the membership degree that it belongs to the superclass, say A, be $\mu_A(e)$. Then $\mu_B(e) \leq \mu_A(e)$.
(b) Assume that a threshold, say β, is given. Then $\mu_B(e) \geq \beta$. Here B, e, and $\mu_B(e)$ are the same as the above.

The subclass B is then a subclass of the superclass A with a membership degree. This membership degree is the minimum of the membership degrees to which these objects belong to the subclass. Here the given threshold is used for a computational threshold to avoid propagating infinitesimal degrees. Formally, we have the definition for fuzzy subclass-superclass relationship.

Let A and B be two (fuzzy) classes with object membership functions μ_A and μ_B, respectively. Let β be a given threshold. We say B is a subclass of A if

$$(\forall e)\ (\beta \leq \mu_B(e) \leq \mu_A(e)).$$

The membership degree that B is a subclass of A should be $\min_{\mu_B(e) \geq \beta}(\mu_B(e))$. Here, e is object instance of A and B in the universe of discourse, and $\mu_A(e)$ and $\mu_B(e)$ are membership degrees of e to A and B, respectively.

It should be noted, however, that in the above-mentioned fuzzy generalization relationship, we assume that classes A and B can only have the second level of fuzziness. It is possible that classes A or B are the classes with membership degree denoted by scalar, namely, with the first level of fuzziness.

Let two classes A and B be A WITH *degree_A* DEGREE and B WITH *degree_B* DEGREE. Here, *degree_A* and *degree_B* are the scalars of membership degree. Let μ_A and μ_B be the object membership functions of A and B, respectively. Then B is a subclass of A if

$$(\forall e)\ (\beta \leq \mu_B(e) \leq \mu_A(e)) \wedge ((\beta \leq degree\_B \leq degree\_A).$$

That means that B is a subclass of A only if, in addition to the condition that the membership degrees of all objects to A and B must be greater than or equal to the given threshold, and the membership degree of any object to A must be greater than or equal to the membership degree of this object to B, the membership degrees of A and B must be greater than or equal to the given threshold, and the membership degree of A must be greater than or equal to the membership degree of B.

Consider a fuzzy superclass A and its fuzzy subclasses B_1, B_2, …, B_n with object membership functions μ_A, μ_{B1}, μ_{B2}, …, and μ_{Bn}, respectively, which may also have the scalars of membership degree *degree_A*, *degree_B*$_1$, *degree_B*$_2$, …, and *degree_B*$_n$, respectively. Then the following relationship is true:

$$(\forall e)\ (\max(\mu_{B1}(e), \mu_{B2}(e), \ldots, \mu_{Bn}(e)) \leq \mu_A(e))$$
$$\wedge\ (\max(degree\_B_1, degree\_B_2, \ldots, degree\_B_n) \leq degree\_A)$$

It can be seen that we can assess fuzzy subclass/superclass relationships by utilizing the inclusion degree of objects to the class. Clearly such an assessment is based on the extensional viewpoint of class. When classes are defined with the intensional viewpoint, there is no object available. Therefore, the method given above cannot be used. At this point, we can use the inclusion degree of a class with respect to another class to determine the relationships between fuzzy subclass and superclass. The basic idea is that, since any object belonging to the subclass should belong to the superclass, the common attribute domains of the superclass should include the common attribute domains of the subclass.

Let *A* and *B* be (fuzzy) classes and the degree that *B* is the subclass of *A* be denoted by μ (*A*, *B*). For a given threshold β, we say *B* is a subclass of *A* if

$$\mu(A, B) \geq \beta.$$

Here, μ (*A*, *B*) is a scalar and used to calculate the inclusion degree of *B* with respect to *A* according to the inclusion degree of the attribute domains of *B* with respect to the attribute domains of *A* as well as the weight of attributes.

To figure out or estimate the inclusion degree of two classes, one needs to know the (fuzzy) attribute domains of the two classes and the weight of the attributes. The problem of evaluating the inclusion degree is outside the scope of the current paper. One can refer to Ma et al. (2004), where the methods for evaluating the inclusion degree of fuzzy attribute domains and further evaluating the inclusion degree of a subclass with respect to the superclass are discussed in detail.

Now let us consider the situation that classes *A* or *B* are the classes with scalars of membership degree, namely, with the first level of fuzziness.

Let two classes *A* and *B* be *A* WITH *degree_A* DEGREE and *B* WITH *degree_B* DEGREE. Here, *degree_A* and *degree_B* are the scalars of membership degree. Then *B* is a subclass of *A* if

$$(\mu(A, B) \geq \beta) \wedge ((\beta \leq degree\_B \leq degree\_A).$$

That means that *B* is a subclass of *A* only if, in addition to the condition that the inclusion degree of *A* with respect to *B* must be greater than or equal to the given threshold, the scalars of membership degree of *A* and *B* must be greater than or equal to the given threshold, and the scalar of the membership degree of *A* must be greater than or equal to the scalar of the membership degree of *B*.

In subclass-superclass hierarchies, a critical issue is multiple inheritance of class. Ambiguity arises when more than one of the superclasses have common attributes, and the subclass does not explicitly declare the class from which the attribute is inherited. Exactly which conflicting attribute in the superclasses is inherited by the subclass depends on their weights to the corresponding superclasses (Liu and Song 2001). Also, it should be noted that in fuzzy multiple inheritance hierarchy, the subclass has different degrees with respect to different superclasses, not being the same as the situation in classical object-oriented databases (Ma et al. 2004).

Several generalizations can be grouped together to form a *class hierarchy* as shown in Fig. 2.3. Figure 2.3 shows a fuzzy generalization relationship, where a *dashed triangular arrowhead* is used to represent a fuzzy generalization relationship. The *disjointness* and *completeness* constraints, which are optional, can be

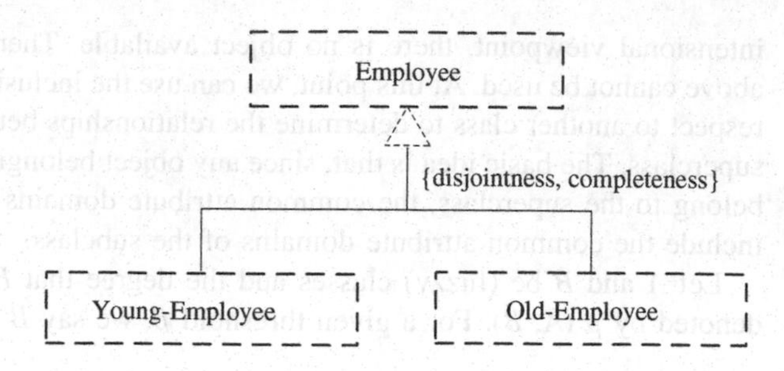

Fig. 2.3 Representation of a fuzzy *generalization* in a fuzzy UML data model

enforced on a class hierarchy. The *disjointness* means that all the specific classes are mutually disjoint, and *completeness* means that the union of the more specific classes completely covers the more general class. That is, the union of object instances of several subclasses completely covers the object instances of the superclass, and the membership degree that any object belongs to the subclass must be less than or equal to the membership degree that it belongs to the superclass.

2.3.4 Fuzzy Aggregation

An aggregation captures a whole-part relationship between a class named aggregate and a group of classes named constituent parts. The constituent parts can exist independently. Aggregate class *Car*, for example, is aggregated by constituent part classes *Engine*, *Interior*, and *Chassis*. Each object of an aggregate can be projected into a set of objects of constituent parts. Formally, let A be an aggregation of constituent parts B_1, B_2, …, and B_n. For $e \in A$, the projection of e to B_i is denoted by $e{\downarrow}_{Bi}$. Then we have

$$(e{\downarrow}_{B1}) \in B_1,\ (e{\downarrow}_{B2}) \in B_2,\ \ldots,\ \text{and}\ (e{\downarrow}_{Bn}) \in B_n.$$

A class aggregated from fuzzy constituent parts must be fuzzy. If the former is still called an aggregate, the aggregation is fuzzy. At this point, a class is an aggregation of constituent parts with membership degree of [0, 1]. We have the following criteria to determine the Fuzzy Aggregation relationship:

(a) For any (fuzzy) object, say e, let the membership degree that it belongs to the aggregate, say A, be $\mu_A(e)$. Also, let the projections of e to the constituent parts, say B_1, B_2, …, and B_n, be $e{\downarrow}_{B1}$, $e{\downarrow}_{B2}$, …, and $e{\downarrow}_{Bn}$. Let the membership degrees that these projections belong to B_1, B_2, …, and B_n be $\mu_{B1}(e{\downarrow}_{B1})$, $\mu_{B2}(e{\downarrow}_{B2})$, …, and $\mu_{Bn}(e{\downarrow}_{Bn})$, respectively. Then $\mu_A(e) \le \mu_{B1}(e{\downarrow}_{B1})$, $\mu_A(e) \le \mu_{B2}(e{\downarrow}_{B2})$, …, and $\mu_A(e) \le \mu_{Bn}(e{\downarrow}_{Bn})$.

(b) Assume that a threshold, say β, is given. Then $\mu_A(e) \ge \beta$. Here, A, e, and $\mu_A(e)$ are the same as the above.

Then A is the aggregate of B_1, B_2, …, and B_n, with the membership degree min (μ_{B1} ($e\downarrow_{B1}$), μ_{B2} ($e\downarrow_{B2}$), …, μ_{Bn} ($e\downarrow_{Bn}$))). It is clear that μ_A (e) cannot have a bigger value than any μ_{B1} ($e\downarrow_{B1}$), μ_{B2} ($e\downarrow_{B2}$), …, and μ_{Bn} ($e\downarrow_{Bn}$). And μ_{B1} ($e\downarrow_{B1}$), μ_{B2} ($e\downarrow_{B2}$), …, and μ_{Bn} ($e\downarrow_{Bn}$) are not aggregated into 1 except that μ_{B1} ($e\downarrow_{B1}$), μ_{B2} ($e\downarrow_{B2}$), …, and μ_{Bn} ($e\downarrow_{Bn}$) are equal to 1. Formally, we have the following definition for the fuzzy aggregation relationship.

Let A be a fuzzy aggregation of fuzzy class sets B_1, B_2, …, and B_n, which object membership functions are μ_A, μ_{B1}, μ_{B2}, …, and μ_{Bn}, respectively. Let β be a given threshold. Then

$$(\forall e)\ (e \in A \wedge \beta \leq \mu_A(e) \leq \min\ (\mu_{B1}(e\downarrow_{B1}), \mu_{B2}(e\downarrow_{B2}), \ldots, \mu_{Bn}(e\downarrow_{Bn}))).$$

That means a fuzzy class A is the aggregate of a of group fuzzy classes B_1, B_2, …, and B_n if, for any (fuzzy) object instance, the scalar of the membership degree that it belongs to class A is less than or equal to the scalar of the member degree to which its projection to any of B_1, B_2, …, and B_n, say B_i ($1 \leq i \leq n$), belongs to class B_i. Meanwhile, for any (fuzzy) object instance, the scalar of the membership degree that it belongs to class A is greater than or equal to the given threshold.

Now let us consider the first level of fuzziness in the above-mentioned classes A, B_1, B_2, …, and B_n; namely, they are the fuzzy classes with membership degrees. Let A WITH *degree_A* DEGREE, B_1 WITH *degree_B*$_1$ DEGREE, B_2 WITH *degree_B*$_2$ DEGREE, …, B_n WITH *degree_B*$_n$ DEGREE be classes. Here *degree_A*, *degree_B*$_1$, *degree_B*$_2$, …, and *degree_B*$_n$ are the scalars of membership degree. Let μ_A, μ_{B1}, μ_{B2}, …, and μ_{Bn} be the object membership functions of A, B_1, B_2, …, and B_n, respectively. Then A is an aggregate of B_1, B_2, …, and B_n if

$$(\forall e)\ (e \in A \wedge \beta \leq \mu_A(e) \leq \min(\mu_{B1}(e\downarrow_{B1}), \mu_{B2}(e\downarrow_{B2}), \ldots, \mu_{Bn}(e\downarrow_{Bn})) \\ \wedge\ degree\_A \leq \min(degree\_B_1, degree\_B_2, \ldots, degree\_B_n)).$$

Here β is a given threshold.

It should be noted that the assessment of fuzzy aggregation relationships given above is based on the extensional viewpoint of class. Clearly, these methods cannot be used if the classes are defined with the intensional viewpoint because there is no object available. In the following, we state how to determine the fuzzy aggregation relationship using the inclusion degree developed in Ma et al. (2004).

Let A be a fuzzy aggregation of fuzzy class sets B_1, B_2, …, and B_n, and β be a given threshold. Also, let the projection of A to B_i be denoted by $A\downarrow_{Bi}$. Then

$$\min(\mu(B_1, A\downarrow_{B1}), \mu(B_2, A\downarrow_{B2}), \ldots, \mu(B_n, A\downarrow_{Bn})) \geq \beta.$$

Being the same as the fuzzy generation, here μ (B_i, $A\downarrow_{Bi}$) ($1 \leq i \leq n$) means the membership degree to which B_i semantically includes $A\downarrow_{Bi}$. The membership degree that A is an aggregation of B_1, B_2, …, and B_n is min ($\mu(B_1, A\downarrow_{B1})$, $\mu(B_2, A\downarrow_{B2}), \ldots, \mu(B_n, A\downarrow_{Bn})$).

Furthermore, the expression above can be extended for the situation that A, B_1, B_2, …, and B_n may have the first level of fuzziness, namely, they may be the fuzzy classes with membership degrees.

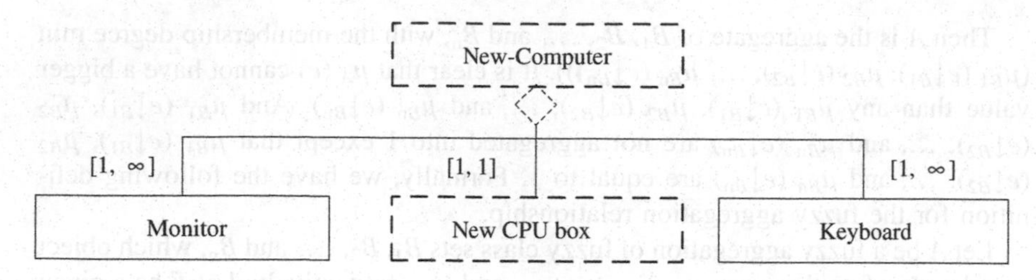

Fig. 2.4 Representation of a *fuzzy aggregation* in a fuzzy UML data model

Let β be a given threshold and A WITH *degree_A* DEGREE, B_1 WITH *degree_B*$_1$ DEGREE, B_2 WITH *degree_B*$_2$ DEGREE, …, B_n WITH *degree_B*$_n$ DEGREE be classes. Then A is an aggregate of B_1, B_2, …, and B_n if

$$\min(\mu(B_1, A\downarrow_{B1}), \mu(B_2, A\downarrow_{B2}), \ldots, \mu(B_n, A\downarrow_{Bn})) \\ \geq \beta \wedge degree\_A \leq \min(degree\_B_1, degree\_B_2, \ldots, degree\_B_n)).$$

Figure 2.4 shows a fuzzy aggregation relationship, where a *dashed open diamond* is used to denote a fuzzy aggregation relationship. Here:

- The class *New CPU box* is a fuzzy class, and thus the class *New-Computer* aggregated by *Monitor*, *New CPU box*, and *Keyboard* is also fuzzy.
- The *multiplicity* $[m_i, n_i]$ specifies that each instance of the aggregate class consists of at least m_i and at most n_i instances of the *i*-th constituent class. For example, a New-Computer may contain at least one Monitor, and possibly more.

2.3.5 Fuzzy Dependency

A *dependency*, which is a relationship between a source class and a target class, denotes that the target class exists dependently on the source class. In addition, the dependency between the source class and the target class is only related to the classes themselves and does not require a set of instances for its meaning. Therefore, the second level of fuzziness and the third level of fuzziness in class do not affect the dependency relationship.

A fuzzy dependency relationship is a dependency relationship with a degree of possibility η as shown in Fig. 2.5, which can be indicated explicitly by the designers or be implied implicitly by the source class based on the fact that the target class is decided by the source class.

Figure 2.5 shows a fuzzy dependency relationship, which is denoted by a *dashed line with an arrowhead*. It is clear that *Employee Dependent* is dependent on fuzzy class *Employee* with membership degree $\eta \in [0, 1]$.

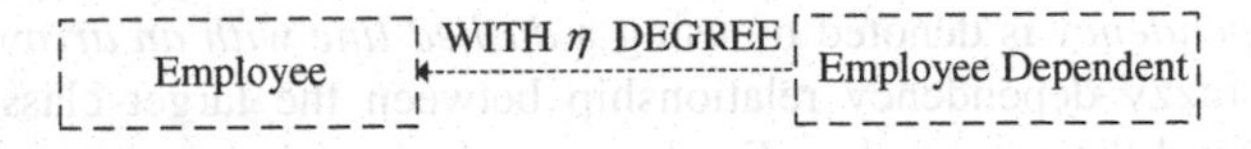

Fig. 2.5 Representation of a *fuzzy dependency* in a fuzzy UML data model

2.3.6 An Example of Fuzzy UML Data Model

Figure 2.6 shows a graphic fuzzy UML data model $\mathcal{F}_{\text{UML}}$, which models the situation at a company. The detailed instruction about the fuzzy UML data model $\mathcal{F}_{\text{UML}}$ is as follows:

- A *fuzzy class* is denoted by using a *dashed-outline rectangle* to differentiate a classical class, e.g., *Old-Employee* as shown in Fig. 2.6. A fuzzy class may contain four parts, i.e., crisp attributes, fuzzy attributes, methods, or an additional attribute u as shown in Fig. 2.6, where:
  - A fuzzy keyword *FUZZY* is appeared in front of an attribute indicating the attribute may take fuzzy values. For example, *FUZZY* Age or *FUZZY* Email. Moreover, the *multiplicity* $[1\ldots\infty]$ of the attribute "*FUZZY* Email $[1\ldots\infty]$" means that each object instance of the fuzzy class *Old-Employee* has at least one email, and possibly more.
  - An additional attribute $u \in [0, 1]$ in a fuzzy class is defined for representing an object membership degree to the fuzzy class.
  - The method *IsDepartment()*: *String* returns a possibility distribution value $\{Department/u_i\}$, which denotes that an *Old-Employee* works in the *Department* with degree $u_i \in [0, 1]$. The type of the parameter is *null*.
- A *fuzzy generalization* is denoted by using a *dashed triangular arrowhead*, e.g., the class *Employee* is a generalization of classes *Young-Employee* and *Old-Employee*, and two classes *Young-Employee* and *Old-Employee* are *disjoint* and *the union of them* completely covers the class *Employee*.

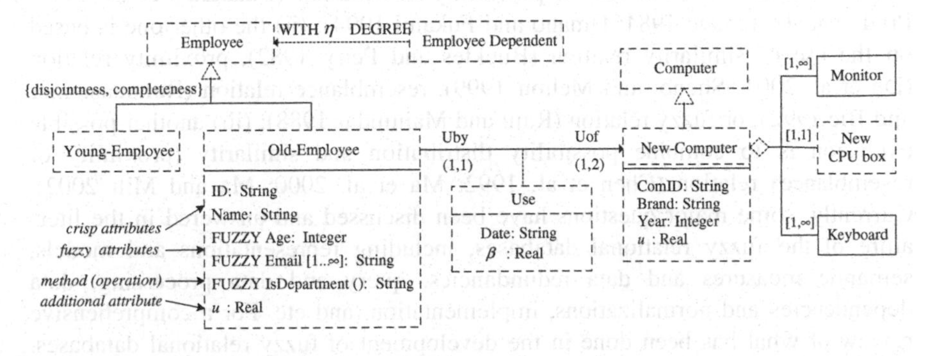

Fig. 2.6 A graphic fuzzy UML data model $\mathcal{F}_{\text{UML}}$

- A *fuzzy dependency* is denoted by using a *dashed line with an arrowhead*, e.g., there is a fuzzy dependency relationship between the target class *Employee Dependent* and the source class *Employee*.
- A *fuzzy association* is denoted by using a *single line with an arrowhead* and the association class is connected with the association by a *dashed-outline*, e.g., *Use* is a fuzzy association class between two classes *Old-Employee* and *New-Computer*. Here:

  – *Date* is an attribute of the association class *Use*, which describes the start date that an *Old-Employee* uses a *New-Computer*.
  – The additional attribute β denotes the degree of possibility that an association relationship occurs in n classes.
  – The participation of a class in a fuzzy association is called a *role* which has a unique name. For example, *Uby* and *Uof*. The cardinality constraints (1, 1) and (1, 2) denote that each *Old-Employee* can use at least 1 and at most 2 *New-Computers* and each *New-Computer* can be used by exactly one *Old-Employee*.
  – A *fuzzy aggregation* is denoted by using a *dashed open diamond*, e.g., the class *New-Computer* is aggregated by *Monitor*, *New CPU box*, and *Keyboard*. The *multiplicity* [m_i, n_i] specifies that each instance of the aggregate class consists of at least m_i and at most n_i instances of the *i*-th constituent class. For example, a *New-Computer* may contain at least one *Monitor*, and possibly more.

2.4 Fuzzy Relational Database Models

In order to manage fuzzy data in the databases, fuzzy set theory has been extensively applied to extend various database models and resulted in numerous contributions, mainly with respect to the popular relational model or to some related form of it. In general, several basic approaches can be classified: (i) one of fuzzy relational databases is based on possibility distribution (Chaudhry et al. 1999; Prade and Testemale 1984; Umano and Fukami 1994); (ii) the other one is based on the use of similarity relation (Buckles and Petry 1982), proximity relation (De et al. 2001; Shenoi and Melton 1999), resemblance relation (Rundensteiner and Bic 1992), or fuzzy relation (Raju and Majumdar 1988); (iii) another possible extension is to combine possibility distribution and similarity (proximity or resemblance) relation (Chen et al. 1992; Ma et al. 2000; Ma and Mili 2002). Currently, some major questions have been discussed and answered in the literature of the fuzzy relational databases, including representations and models, semantic measures and data redundancies, query and data processing, data dependencies and normalizations, implementation, and etc. For a comprehensive review of what has been done in the development of fuzzy relational databases, please refer to Chen (1999), Ma (2005b), Ma and Yan (2008), Petry (1996), and Yazici and George (1999). In this section, we briefly introduce some basic notions of fuzzy relational databases based on possibility distributions.

2.4.1 Fuzzy Relational Database Models

Basically, a *fuzzy relational database* (*FRDB*) is based on the notions of *fuzzy relational schema*, *fuzzy relational instance*, *tuple*, *key*, and *constraints*, which are introduced briefly as follows:

- A *fuzzy relational database* consists of a set of *fuzzy relational schemas* and a set of *fuzzy relational instances* (i.e., simply *fuzzy relations*).
- The set of *fuzzy relational schemas* specifies the structure of the data held in a database. A fuzzy relational schema consists of a fixed set of attributes with associated domains. The information of a domain is implied in forms of schemas, attributes, keys, and referential integrity constraints.
- The set of *fuzzy relations*, which is considered to be an instance of the set of fuzzy relation schemas, reflects the real state of a database. Formally, a fuzzy relation is a two-dimensional array of rows and columns, where each column represents an attribute and each row represents a tuple.
- Each *tuple* in a table denotes an *individual* in the real world identified uniquely by primary key, and a foreign key is used to ensure the data integrity of a table. A column (or columns) in a table that makes a row in the table distinguishable from other rows in the same table is called the *primary key*. A column (or columns) in a table that draws its values from a primary key column in another table is called the *foreign key*. As is generally assumed in the literature, we assume that the primary key attribute is always crisp and all fuzzy relations are in the third normal form.
- An *integrity constraint* in a schema is a predicate over relations expressing a constraint; by far the most used integrity constraint is the referential integrity constraint. A *referential integrity constraint* involves two sets of attributes S_1 and S_2 in two relations R_1 and R_2, such that one of the sets (say S_1) is a key for one of the relations (called primary key). The other set is called a foreign key if R_2 $[S_2]$ is a subset of R_1 $[S_1]$. Referential integrity constraints are the glue that holds the relations in a database together.

In summary, in a fuzzy relational database, the structure of the data is represented by a set of *fuzzy relational schemas*, and data are stored in *fuzzy relations* (i.e., *tables*). Each table contains *rows* (i.e., *tuples*) and *columns* (i.e., *attributes*). Each tuple is identified uniquely by the *primary key*. The relationships among relations are represented by the referential integrity constraints, i.e., *foreign keys*. Moreover, here, two types of fuzziness are considered in fuzzy relational databases, one is the fuzziness of attribute values (i.e., attributes may be fuzzy), which may be represented by possibility distributions; another is the fuzziness of a tuple being a member of the corresponding relation, which is represented by a membership degree associated with the tuple.

Formally, a fuzzy relational database *FRDB* = <*FS*, *FR*> consists of a set of fuzzy relational schemas *FS* and a set of fuzzy relations *FR*, where:

- Each *fuzzy relational schema FS* can be represented formally as *FR* (A_1/D_1, A_2/D_2, …, A_n/D_n, μ_{FR}/D_{FR}), which denotes that a fuzzy relation *FR* has attributes A_1, A_2, …, A_n and μ_{FR} with associated data types D_1, D_2, …, D_n and D_{FR}. Here, μ_{FR} is an additional attribute for representing the membership degree of a tuple to the fuzzy relation.
- Each *fuzzy relation FR* on a fuzzy relational schema *FR* (A_1/D_1, A_2/D_2, …, A_n/D_n, μ_{FR}/D_{FR}) is a subset of the Cartesian product of *Dom* (A_1) × *Dom* (A_2) × ⋯ × *Dom* (A_n) × *Dom* (μ_{FR}), where *Dom* (A_i) may be a fuzzy subset or even a set of fuzzy subset and *Dom* (μ_{FR}) ∈ (0, 1]. Here, *Dom* (A_i) denotes the domain of attribute A_i, and each element of the domain satisfies the constraint of the datatype D_i. Formally, each *tuple* in *FR* has the form $t = <\pi_{A1}, \pi_{A2}, \ldots, \pi_{Ai}, \ldots, \pi_{An}, \mu_{FR}>$, where the value of an attribute A_i may be represented by a possibility distribution π_{Ai}, and $\mu_{FR} \in (0, 1]$.

Moreover, a resemblance relation *Res* on *Dom* (A_i) is a mapping: *Dom* (A_i) × *Dom* (A_i) → [0, 1] such that (i) for all *x* in Dom (A_i), *Res* (*x*, *x*) = 1 (reflexivity) (ii) for all *x*, *y* in Dom (A_i), *Res* (*x*, *y*) = *Res* (*y*, *x*) (symmetry).

To provide the intuition on the fuzzy relational database we show an example. The following gives a fuzzy relational database modeling parts of the reality at a company, including fuzzy relational schemas in Table 2.1 and fuzzy relations in Table 2.2. The detailed introduction is as follows:

- The attribute underlined stands for primary key *PK*. The foreign key (*FK*) is followed by the parenthesized relation called referenced relation. A relation can have several candidate keys from which one primary key, denoted *PK*, is chosen.
- An '*f*' next to an attribute means that the attribute is fuzzy.
- In Table 2.1, there are the inheritance relationships *Chief-Leader* "is-a" *Leader* and *Young-Employee* "is-a" *Employee*. There is a *1-many relationship* between *Department* and *Young-Employee*. The relation *Supervise* is a *relationship relation*, and there is *many–many relationship* between *Chief-Leader* and *Young-Employee*.
- Note that, a relation is different from a relationship. A relation is essentially a table, and a relationship is a way to correlate, join, or associate the two tables.

2.4.2 Fuzzy Relational Algebra Operations

In the following, we introduce the fuzzy relational operations in relational algebra (Ma and Mili 2002), which can be used to query and update fuzzy relational

Table 2.1 The fuzzy relational schemas of a fuzzy relational database

Relation name	Attribute and datatype	Foreign key and referenced relation
Leader	leaID (String), lNumber (String), μ_{FR} (Real)	No
Employee	empID (String), eNumber (String), μ_{FR} (Real)	No
Chief-Leader	leaID (String), clName (String), f_clAge (Integer), μ_{FR} (Real)	leaID (Leader (leaID))
Young-Employee	empID (String), yeName (String), f_yeAge (*Integer*), f_yeSalary (Integer), dep_ID (String), μ_{FR} (Real)	empID (Employee (empID)) dept_ID (Department (depID))
Supervise	supID (String), lea_ID (String), emp_ID (String), μ_{FR} (Real)	lea_ID (Chief-Leader (leaID)) emp_ID (Young-Employee (empID))
Department	depID (String), dName (String), μ_{FR} (Real)	No

databases. It will be shown that the fuzzy relational operations, being similar to the conventional relational databases, are complete and sound. Before introducing the fuzzy relational algebra operations, we will first introduce the notion of the semantic measure of fuzzy data.

The semantics of a fuzzy data represented by possibility distribution corresponds to the semantic space and the semantic relationship between two fuzzy data can be described by the relationship between their semantic spaces (Rundensteiner et al. 1989). The semantic inclusion degree is then employed to measure semantic inclusion and thus measure semantic equivalence of fuzzy data.

Let π_A and π_B be two fuzzy data, and their semantic spaces be SS (π_A) and SS (π_B), respectively. Let SID (π_A, π_B) denotes the degree that π_A semantically includes π_B. Then

$$\text{SID}\ (\pi_A, \pi_B) = (\text{SS}\ (\pi_B) \cap \text{SS}\ (\pi_A))/\text{SS}\ (\pi_B)$$

For two fuzzy data π_A and π_B, the meaning of SID (π_A, π_B) is the percentage of the semantic space of π_B which is wholly included in the semantic space of π_A. Therefore, the concept of equivalence degree can be easily drawn as follows.

Let π_A and π_B be two fuzzy data and SID (π_A, π_B) be the degree that π_A semantically includes π_B. Let SE (π_A, π_B) denote the degree that π_A and π_B are equivalent to each other. Then

$$\text{SE}\ (\pi_A, \pi_B) = min(\text{SID}\ (\pi_A, \pi_B),\ \text{SID}\ (\pi_B, \pi_A))$$

On this basis, based on possibility distribution and resemblance relation, the notion for calculating the semantic inclusion degree of two fuzzy data is given as follows.

Let $U = \{u_1, u_2, \ldots, u_n\}$ be the universe of discourse. Let π_A and π_B be two fuzzy data on U based on possibility distribution and π_A (u_i), $u_i \in U$, denote the

Table 2.2 The fuzzy relations of a fuzzy relational database

Leader

leaID	*lNumber*	μ_{FR}
L001	001	0.7
L002	002	0.9
L003	003	0.8

Employee

empID	*eNumber*	μ_{FR}
E001	001	0.8
E002	002	0.9

Chief-Leader

leaID	*clName*	*f_clAge*	μ_{FR}
L001	Chris	{35/0.8, 39/0.9}	0.65
L003	Billy	37	0.7

Young-Employee

empID	*yeName*	*f_yeAge*	*f_yeSalary*	*dep_ID*	μ_{FR}
E001	John	{24/0.7, 25/0.9}	{2000/0.3, 3000/0.4}	D001	0.75
E002	Mary	23	{4000/0.5, 4500/0.7, 5000/1.0}	D003	0.85

Department

depID	*dName*	μ_{FR}
D001	HR	0.8
D002	Finance	0.9
D003	Sales	0.7

Supervise

supID	*lea_ID*	*emp_ID*	μ_{FR}
S001	L001	E001	0.78
S002	L001	E002	0.8
S002	L003	E002	0.9

possibility that u_i is true. Let *Res* be a resemblance relation on domain *U*, and α $(1 \leq \alpha \leq n)$ be a threshold corresponding to *Res*. SID (π_A, π_B) is then defined by

$$\mathrm{SID}_\alpha(\pi_A, \pi_B) = \sum_{i=1}^{n} \min_{u_i, u_j \in U \text{ and } \mathrm{Res}_U(u_i, u_j) \geq \alpha} (\pi_B(u_i), \pi_A(u_j)) \Big/ \sum_{i=1}^{n} \pi_B(u_i)$$

The notion of the semantic equivalence degree of attribute values can be extended to the semantic equivalence degree of tuples. Let $t_i = <a_{i1}, a_{i2}, \ldots, a_{in}>$ and $t_j = <a_{j1}, a_{j2}, \ldots, a_{jn}>$ be two tuples in fuzzy relational instance *r* over schema *FR* $(A_1, A_2, \ldots, A_n)$. The semantic equivalence degree of tuples t_i

and t_j is denoted SE (t_i, t_j) = min {SE (t_i [A_1], t_j [A_1]), SE (t_i [A_2], t_j [A_2]), …, SE (t_i [A_n], t_j [A_n])}.

Following the semantic inclusion degree of fuzzy data, two kinds of fuzzy data redundancies can be classified and evaluated, which are inclusion redundancy and equivalence redundancy. Being different from the classical set theory, the condition SS (A) $\supseteq$ SS (B) or SS (A) $\subseteq$ SS (B) is essentially the particular case of fuzzy data due to the fuzziness of the data. In general, the threshold should be considered when evaluating the semantic relationship between two fuzzy data.

Let π_A and π_B be two fuzzy data and β be a threshold. If SID (π_A, π_B) $\geq \beta$, π_B is said to be *inclusively* redundant to π_A. If SE (π_A, π_B) $\geq \beta$, it is said that π_A and π_B are *equivalently* redundant to each other.

It is clear that equivalence redundancy of fuzzy data is a particular case of inclusion redundancy of fuzzy data. Considering the effect of resemblance relation in evaluation of semantic inclusion degree and equivalence degree, two fuzzy data π_A and π_B are considered equivalently α-β-redundant if and only if SE_α (π_A, π_B) $\geq \beta$. If SID_α (π_A, π_B) $\geq \beta$ and SID_α (π_B, π_A) $< \beta$, π_B are inclusively α-β-redundant to π_A.

If π_A and π_B are inclusively redundant or equivalently redundant, the removal of redundancy can be achieved by merging π_A and π_B and producing a new fuzzy data π_C. Following Zadeh's extension principle (Zadeh 1975), the operation with an infix operator "θ" on π_A and π_B can be defined as follows.

$$\begin{aligned}\pi_A\,\theta\,\pi_B &= \{\pi_A(u_i)/u_i \mid u_i \in U \wedge 1 \leq i \leq n\}\,\theta\,\{\pi_B(v_j)/v_j \mid v_j \in U \wedge 1 \leq j \leq n\}\\ &= \{\max(\min(\pi_A(u_i), \pi_B(v_j))/u_i\theta v_j) \mid u_i, v_j \in U \wedge 1 \leq i, j \leq n\}\end{aligned}$$

Assume that π_A and π_B are α-β-redundant to each other, the elimination of duplicate could be achieved by merging π_A and π_B and producing a new fuzzy data π_C, where π_A, π_B and π_C are three fuzzy data on $U = \{u_1, u_2, \ldots, u_n\}$ based on possibility distribution and there is a resemblance relation Res_U on U. Then the following three merging operations are defined:

(a) $\pi_C = \pi_A \cup_f \pi_B = \{\pi_C(w)/w \mid (\exists\, \pi_A(u_i)/u_i)\, (\exists\, \pi_B(v_j)/v_j)\, (\pi_C(w) = \max(\pi_A(u_i), \pi_B(v_j)) \wedge (w = u_i \mid \pi_{C(w)} = \pi_{A(ui)} \vee w = v_j \mid \pi_{C(w)} = \pi_{B(vj)}) \wedge Res_U(u_i, v_j) \geq \alpha \wedge u_i, v_j \in U \wedge 1 \leq i, j \leq n) \vee (\exists\, \pi_A(u_i)/u_i)\, (\forall\, \pi_B(v_j)/v_j)\, (\pi_C(w) = \pi_A(u_i) \wedge w = u_i \wedge Res_U(u_i, v_j) \geq \alpha \wedge u_i, v_j \in U \wedge 1 \leq i, j \leq n) \vee (\exists\, \pi_B(v_j)/(v_j)\, (\forall\, \pi_A(u_i)/u_i)\, (\pi_C(w) = \pi_B(v_j) \wedge w = v_j \wedge Res_U(u_i, v_j) \geq \alpha \wedge u_i, v_j \in U \wedge 1 \leq i, j \leq n)\}$,

(b) $\pi_C = \pi_A -_f \pi_B = \{\pi_C(w)/w \mid (\exists\, \pi_A(u_i)/u_i)\, (\exists\, \pi_B(v_j)/v_j)\, (\pi_C(w) = \max(\pi_A(u_i) - \pi_B(v_j), 0) \wedge w = u_i \wedge Res_U(u_i, v_j) \geq \alpha \wedge u_i, v_j \in U \wedge 1 \leq i, j \leq n) \vee (\exists\, \pi_A(u_i)/u_i)\, (\forall\, \pi_B(v_j)/v_j)\, (\pi_C(w) = \pi_A(u_i) \wedge w = u_i \wedge Res_U(u_i, v_j) < \alpha \wedge u_i, v_j \in U \wedge 1 \leq i, j \leq n)\}$, and

(c) $\pi_C = \pi_A \cap_f \pi_B = \{\pi_C(w)/w \mid (\exists\, \pi_A(u_i)/u_i)\, (\exists\, \pi_B(v_j)/v_j)\, (\pi_C(w) = \min(\pi_A(u_i), \pi_B(v_j)) \wedge (w = u_i \mid \pi_{C(w)} = \pi_{A(ui)} \vee w = v_j \mid \pi_{C(w)} = \pi_{B(vj)}) \wedge Res_U(u_i, v_j) \geq \alpha \wedge u_i, v_j \in U \wedge 1 \leq i, j \leq n)\}$.

Let $\pi_A = \{1.0/a, 0.95/b, 0.9/c, 0.2/f\}$ and $\pi_B = \{0.95/a, 0.9/b, 1.0/d, 0.3/e\}$ be two fuzzy data on domain $U = \{a, b, c, d, e, f\}$. Res_U is a resemblance relation on U given in the following, where threshold $\alpha = 0.9$.

Res_U	a	b	c	d	e	f
a	1.0	0.1	0.4	0.3	0.1	0.1
b		1.0	0.2	0.3	0.2	0.2
c			1.0	0.95	0.5	0.3
d				1.0	0.3	0.1
e					1.0	0.4
f						1.0

Then:

$SID_\alpha(\pi_A, \pi_B) = (0.95 + 0.9 + 0.9)/(0.95 + 0.9 + 1.0 + 0.3) = 0.873$,
$SID_\alpha(\pi_B, \pi_A) = (0.95 + 0.9 + 0.9)/(1.0 + 0.95 + 0.9 + 0.2) = 0.902$, and thus
$SE_\alpha(\pi_A, \pi_B) = \min(SID(\pi_A, \pi_B), SID(\pi_B, \pi_A)) = \min(0.873, 0.902) = 0.873$.

If a threshold $\beta = 0.85$ is given, π_A and π_B are considered redundant to each other. Utilizing the merging operations above, one has the following results.

$$\pi_A \cup_f \pi_B = \{1.0/a, 0.95/b, 1.0/d, 0.3/e, 0.2/f\},$$
$$\pi_A -_f \pi_B = \{0.05/a, 0.05/b, 0.2/f\}, \text{ and}$$
$$\pi_A \cap_f \pi_B = \{0.95/a, 0.9/b, 0.9/c\}.$$

The processing of fuzzy value redundancy can be extended to that of fuzzy tuple redundancy. In a similar way, fuzzy tuple redundancy can be classified into inclusion redundancy and equivalence redundancy of tuples.

Let r be a fuzzy relation on the schema $FR(A_1, A_2, \ldots, A_n)$. Let $t = (\pi_{A1}, \pi_{A1}, \ldots, \pi_{An})$ and $t' = (\pi_{A1}', \pi_{A2}', \ldots, \pi_{An}')$ be two tuples in r. Let $\alpha \in [0, 1]$ and $\beta \in [0, 1]$ be two thresholds. The tuple t' is inclusively α-β-redundant to t if and only if $\min(SID_\alpha(\pi_{Ai}, \pi_{Ai}')) \geq \beta$ holds true $(1 \leq i \leq n)$. The tuples t and t′ are equivalently α-β-redundant if and only if $\min(SE_\alpha(\pi_{Ai}, \pi_{Ai}')) \geq \beta$ holds $(1 \leq i \leq n)$.

Based on the notion of the semantic measure of fuzzy data above, in the following we introduce several common fuzzy relational algebra operations.

Union Let r and s be two union-compatible fuzzy relations on the scheme $FR(A_1, A_2, \ldots, A_n)$. Let $\alpha = \{\alpha_i \mid \alpha_i \in [0, 1] \wedge 1 < i \leq n\}$ be the threshold of the resemblance relations on attribute domains and $\beta \in [0, 1]$ be a given threshold. Then the union of these two relations is defined as follows. It is clear that fuzzy union is essentially a α-β-union.

$$r \cup s = \{t \mid (\forall y)\,(y \in s \wedge t \in r \Rightarrow SE_\alpha(t, y) < \beta) \vee (\forall x)\,(x \in r \wedge t \in s \Rightarrow SE_\alpha(t, x) < \beta) \vee ((\exists x)\,(\exists y)\,(x \in r \wedge y \in s \Rightarrow SE_\alpha(x, y) \geq \beta \wedge t = x \cup_f y)\}$$

Difference Let r and s be the same as the above. Their difference is defined as follows.

$$r - s = \{t \mid (\forall y)\,(y \in s \wedge t \in r \wedge \mathrm{SE}_\alpha(t, y) < \beta) \vee ((\exists x)\,(\exists y)\,(x \in r \wedge y \in s \wedge \mathrm{SE}_\alpha(x, y) \geq \beta \wedge t = x -_f y))\}$$

Cartesian Product The Cartesian product of fuzzy relations is the same as one under classical relational databases. Let r and s be two fuzzy relations on schema R and S, respectively. Then

$$r \times s = \{t[R \cup S] \mid (\forall x)\,(\forall y)\,(x \in r \wedge y \in s \wedge t[R] = x[R] \wedge t[S] = y[S])\}$$

Selection Let r (R) be a fuzzy relation and P be a predicate denoted selection condition. In classical relational databases, a predicate is formed through combining the basic clause $X\ \theta\ Y$ as operands with operators $\neg$, $\wedge$, and $\vee$, where $\theta \in \{>, <, =, \neq, \geq, \leq\}$, and X and Y may be constants, attributes, or expressions which are formed through combining constants, attributes or expressions with arithmetic operations. Under a fuzzy relational database environment, the predicate P may be fuzzy, denoted P_f, to implement fuzzy query for fuzzy databases. In P_f, the constants and attributes may be fuzzy, so the expressions may also be fuzzy. The evaluation of a fuzzy expression can be conducted by using Zadeh's extension principle. Based on the same consideration, the "θ" should be fuzzy comparison operations $\succ_\beta$, $\prec_\beta$, $\succcurlyeq_\beta$, $\preccurlyeq_\beta$, $\approx_\beta$, and $\not\approx_\beta$, in P_f, where β is a threshold. Let π_A and π_B be two fuzzy data over $U = \{u_1, u_2, \ldots, u_n\}$ and α be the threshold of the resemblance relation on U. Then

(a) $\pi_A \approx_\beta \pi_B$ if $\mathrm{SE}_\alpha(\pi_A, \pi_B) \geq \beta$,
(b) $\pi_A \not\approx_\beta \pi_B$ if $\mathrm{SE}_\alpha(\pi_A, \pi_B) < \beta$,
(c) $\pi_A \succ_\beta \pi_B$ if $\pi_A \not\approx_\beta \pi_B$ and max (supp (π_A)) > max (supp (π_B)),
(d) $\pi_A \succcurlyeq_\beta \pi_B$ if $\pi_A \approx_\beta \pi_B$ or $\pi_A \succ_\beta \pi_B$,
(e) $\pi_A \prec_\beta \pi_B$ if $\pi_A \not\approx_\beta \pi_B$ and max (supp (π_A)) < max (supp (π_B)), and
(f) $\pi_A \preccurlyeq_\beta \pi_B$ if $\pi_A \not\approx_\beta \pi_B$ or $\pi_A \prec_\beta \pi_B$.

Then the selection on r for P_f is defined as follows.

$$\sigma_{Pf}(r) = \{t \mid t \in r \wedge P_f(t)\}$$

Projection Let r (R) be a fuzzy relation and attribute subset $S \subset R$. The projection of r on S is defined as follows.

$$\Pi_S(r) = \{t \mid (\forall x)\,(x \in r \wedge t = \cup_f x)\}$$

The five operations above are called primitive operations in relational databases. There are three additional operation intersection, join, and division, which can be defined by the primitive operations.

Intersection Let r and s be two union-compatible fuzzy relations. Then fuzzy intersection of these two relations can be defined in terms of fuzzy difference operation as:

$$r \cap s = r - (r - s).$$

Join Let r (R) and s (S) be any two fuzzy relations. P_f is a conditional predicate in the form of $A\ \theta\ B$, where $\theta \in \{\succ_\beta, \prec_\beta, \succcurlyeq_\beta, \preccurlyeq_\beta, \approx_\beta, \not\approx_\beta\}$, $A \in R$, and $B \in S$. Then fuzzy join of these two relations can be defined in terms of fuzzy selection operation as:

$$r \bowtie_{Pf} s = \sigma_{Pf}(r \times s).$$

when attributes A and B are identical and "θ" takes $\approx_\beta$, fuzzy join becomes fuzzy natural join, denoted $r \bowtie s$. Being the special case of fuzzy join, fuzzy natural join can be evaluated with the definition above. In the following, the definition of fuzzy natural join is given directly. Let $Q = R \cap S$. Then

$$r \bowtie s = \{t[(R-Q) \cup S] \mid (\exists x)\,(\exists y)\,(x \in r \wedge y \in s \wedge \mathrm{SE}_a(x[Q], y[Q]) \geq \beta \wedge t[R-Q] = x[R-Q] \wedge t[S-Q] = y[S-Q] \wedge t[Q] = x[Q] \cap_f y[Q])\}$$

Division Division, referred to quotient operation sometimes, is used to find out the sub-relation $r \div s$ of a relation r, containing sub-tuples of r which have for complements in r all the tuples of a relation s. In classical relational databases, the division operation is defined by

$$r \div s = \{t \mid (\forall u)\,(u \in s \wedge (t, u) \in r)\},$$

where u is a tuple of s and t a sub-tuple of r such that (t, u) is a tuple of r. Let r (R) and s (S) be two fuzzy relations, where $S \subset R$. Let $Q = R - S$. Then the fuzzy division of r and s can be defined as:

$$r \div s = \Pi_Q(r) - \Pi_Q(Q(r) \times s - r).$$

The rename and outerunion are the other relational operations in addition to the eight operation defined above. Their definitions are given as follows.

Rename This operation is used to change the names of some attributes of a relation. Let r (R) be a fuzzy relation, and let A and B be two attributes satisfying $A \in R$ and $B \notin R$, where A and B have the same domain. Let $S = (R - \{A\}) \cup \{B\}$. Then r with A renamed to B is defined as

$$\rho_{A \leftarrow B}(r) = \{t[S] \mid (\forall y)\,(y \in r \wedge t[S-B] = y[R-A] \wedge t[B] = y[A])\}.$$

Outerunion The common union operation requires two source relations be union-compatible. In order to integrate heterogeneous multiple relations, outer-union operation has widely been used in relational databases. Here, the definition of fuzzy outerunion operation is given so that heterogeneous fuzzy data resources can be integrated.

Let r (K, A, X) and s (K, A, Y) be two fuzzy relations, where K is primary key. The outerunion of r and s, denoted by $r\ \tilde{\cup}\ s$, is defined as

$r\ \tilde{\cup}\ s = \{t\ [KAXY] \mid (\exists\ u)\ (\exists\ v)\ (u \in r \wedge v \in s \wedge t\ [K] = u\ [K] = v\ [K] \wedge (\forall\ S)\ (S \in A \wedge t\ [S] = u\ [S] \cup_f v\ [S]) \wedge t\ [X] = u\ [X] \wedge t\ [Y]) = v\ [Y]) \vee (\exists\ u)\ (\forall\ v)$

$(u \in r \wedge v \in s \wedge t\,[K] = u\,[K] \wedge t\,[A] = u\,[A] \wedge t\,[X] = u\,[X] \wedge t\,[Y] = \varphi \wedge v\,[K] \neq t\,[K]) \vee (\exists\, v)\,(\forall\, u)\,(v \in s \wedge u \in r \wedge t\,[K] = v\,[K] \wedge t\,[A] = v\,[A] \wedge t\,[X] = v\,[Y] \wedge t\,[X] = \varphi \wedge u\,[K] \neq t\,[K])\}$.

Now we discuss several properties of the fuzzy relational algebra. Being similar to the conventional relational databases, the proposed fuzzy relational algebra is sound. In other words, it is closed. It means that the results of all operations are valid relations. In detail, the result relations produced by fuzzy relational operations satisfy the following three criteria:

(a) the attribute values must come from an appropriate attribute domain,
(b) there are no duplicate tuples in a relation, and
(c) the relation must be a finite set of tuples.

Projection, division, and selection take out a part from the source relation in either column direction or row direction. Because the attribute values in source relation must belong to the appropriate attribute domain, the attribute values in these three result relations must come from the appropriate attribute domain. Union, difference, and intersection operations are conducted under union-compatible condition, which satisfies the first criterion. In join and Cartesian product, the attribute values in result relations come from two source relations, respectively, and they must be within the appropriate attribute domains. It is clear that rename operation satisfies the first criterion. As to outerunion operation, it is similar to natural join and thus the first criterion can also be satisfied.

For selection and rename, if there are no redundant tuples in ordinary relation, there are no redundant tuples in the result relations. There exist no redundant tuples in the result relations of union, difference, intersection, join, Cartesian product, division, projection, and outerunion. This can be ensured by the definitions of those operations because the removal of redundancies has been considered. Therefore, the second criterion is satisfied.

Now let us look at the satisfactory situation of the third criterion. Let r and s be two fuzzy relations, and let $|r|$ and $|s|$ denote the numbers of tuples in r and s, respectively. It is easy to see that $0 \leq |\sigma_{Pf}(r)| \leq |r|$ for fuzzy selection. This implies that when no any tuple in r satisfies the selection condition, the tuple number in the result relation is 0, and that when all tuples in r satisfy the selection condition, one obtains $|\sigma_{Pf}(r)| = |r|$. When part of the tuples in r satisfy the selection condition, $|\sigma_{Pf}(r)|$ must be greater than 0 and less than $|r|$. For projection $\Pi_S\,(r)$, if all tuples in r are redundant after projecting, then $|\Pi_S(r)| = 1$; if there is not any redundancy in r after projecting, then $|\Pi_S(r)| = |r|$. In the other situations, $|\Pi_S(r)|$ must be greater than 1 and less than $|r|$, i.e., $1 \leq |\Pi_S(r)| \leq |r|$. Additionally, $|r \cup s|$ must not be greater than $|r| + |s|$, $|r - s|$, $|r \cap s|$ and $|r \div s|$ must not be greater than $|r|$, and $|r \bowtie_{Pf} s|$ and $|r \times s|$ must not be greater than $|r| \times |s|$. In addition, $|\rho_{A \leftarrow B}(r)| = |r|$ and $|\, \tilde{\cup} * \sigma| \leq |r| + |s|$ and $|r\, \tilde{\cup}\, s| \leq |r| + |s|$. Since the number of tuples in the result relation is closely related with the source relations and the source relations are finite, the result relations must be finite.

In addition, fuzzy set operations in relational algebra have the same properties as those of classical set operations. Let r, s, and u be three union-compatible fuzzy relations. Then

(a) $r \cup s = s \cup r$ and $r \cap s = s \cap r$, (commutativity)
(b) $r \cup r = r$ and $r \cap r = r$, (idempotence)
(c) $r \cap (r \cup s) = r$ and $r \cup (r \cap s) = r$, (absorption)
(d) $(r \cup s) \cup u = r \cup (s \cup u)$ and $(r \cap s) \cap u = r \cap (s \cap u)$, (associativity)
(e) $r \cap (s \cup u) = (r \cap s) \cup (r \cap s)$ and $r \cup (s \cap u) = (r \cup s) \cap (r \cup s)$, (distributivity)
(f) $r \cup s = r \cup (s - r)$ and $r \cap s = r - (r - s)$.

The following properties are also held in fuzzy operations in relational algebra. Let r and s be two fuzzy relations on schema R and u be a fuzzy relation on schema Q. Let P_f be a selection predicate involving attributes of R. Then

(a) $u \bowtie (r \cup s) = (u \bowtie r) \cup (u \bowtie s)$ and $u \bowtie (r - s) = (u \bowtie r) - (u \bowtie s)$,
(b) $\sigma_{Pf} (r \cup s) = \sigma_{Pf} (r) \cup \sigma_{Pf} (s)$ and $\sigma_{Pf} (r - s) = \sigma_{Pf} (r) - \sigma_{Pf} (s)$, and
(c) $\sigma_{Pf} (u \bowtie r) = u \bowtie \sigma_{Pf} (r)$.

These properties can be proven by the definitions of fuzzy operations in relational algebra.

2.5 Fuzzy Object-Oriented Database Models

In some real-world applications (e.g., CAD/CAM, multimedia and GIS), they characteristically require the modeling and manipulation of complex objects and semantic relationships. It has been proved that the object-oriented paradigm lends itself extremely well to the requirements. Since classical relational database model and its extension of fuzziness do not satisfy the need of modeling complex objects with imprecision and uncertainty, currently many researches have been concentrated on fuzzy object-oriented database models in order to deal with complex objects and uncertain data together. Zicari and Milano (1990) first introduced incomplete information, namely, null values, where incomplete schema and incomplete objects can be distinguished. From then on, the incorporation of imprecise and uncertain information in object-oriented databases has increasingly received attention. A fuzzy object-oriented database model was defined in Bordogna and Pasi (2001) based on the extension of a graphs-based object model. Based on similarity relationship, uncertainty management issues in the object-oriented database model were discussed in George et al. (1996). Based on possibility theory, vagueness and uncertainty were represented in class hierarchies in Dubois et al. (1991). In more detail, also based on possibility distribution theory, Ma et al. (2004) introduced fuzzy object-oriented database models, some major notions such as objects, classes, objects-classes relationships and subclass/superclass relationships were extended under fuzzy information environment.

Moreover, other fuzzy extensions of object-oriented databases were developed. In Marín et al. (2000, 2001), fuzzy types were added into fuzzy object-oriented databases to manage vague structures. The fuzzy relationships and fuzzy behavior in fuzzy object-oriented database models were discussed in Cross (2001), Gyseghem and Caluwe (1995). Several intelligent fuzzy object-oriented database architectures were proposed in Koyuncu and Yazici (2003), Ndouse (1997), and Ozgur et al. (2009). The other efforts on how to model fuzziness and uncertainty in object-oriented database models were done in Lee et al. (1999), Majumdar et al. (2002), and Umano et al. (1998). The fuzzy and probabilistic object bases (Cao and Rossiter 2003; Nam et al. 2007), fuzzy deductive object-oriented databases (Yazici and Koyuncu 1997), and fuzzy object-relational databases (Cubero et al. 2004) were also developed. In addition, an object-oriented database modeling technique was proposed based on the level-2 fuzzy sets in de Tré and de Caluwe (2003), where the authors also discussed how the object Data Management Group (ODMG) data model can be generalized to handle fuzzy data in a more advantageous way. Also, the other efforts have been paid on the establishment of consistent framework for a fuzzy object-oriented database model based on the standard for the ODMG object data model (Cross et al. 1997). More recently, how to manage fuzziness on conventional object-oriented platforms was introduced in Berzal et al. (2007). Yan and Ma (2012) proposed the approach for the comparison of entity with fuzzy data types in fuzzy object-oriented databases. Yan et al. (2012) investigated the algebraic operations in fuzzy object-oriented databases, and discussed fuzzy querying strategies and gave the form of SQL-like fuzzy querying for the fuzzy object-oriented databases. In the section, the basic notions of fuzzy object-oriented database (*FOODB*) models, including fuzzy object, fuzzy class, fuzzy inheritance, and algebraic operations are introduced.

2.5.1 Fuzzy Objects

Objects model real-world entities or abstract concepts. Objects have properties that may be attributes of the object itself or relationships also known as associations between the object and one or more other objects. An object is fuzzy because of a lack of information. For example, an object representing a part in preliminary design for certain will also be made of *stainless steel*, *moulded steel*, or *alloy steel* (each of them may be connected with a possibility, say, 0.7, 0.5 and 0.9, respectively). Formally, objects that have at least one attribute whose value is a fuzzy set are fuzzy objects.

2.5.2 Fuzzy Classes

The fuzzy classes in fuzzy object-oriented databases are similar to the notion of the fuzzy classes in fuzzy UML data models as introduced in Sect. 2.3.

The objects having the same properties are gathered into classes that are organized into hierarchies. Theoretically, a class can be considered from two different viewpoints (Dubois et al. 1991): (a) an *extensional* class, where the class is defined by the list of its object instances, and (b) an *intensional* class, where the class is defined by a set of attributes and their admissible values. In addition, a subclass defined from its superclass by means of inheritance mechanism in the object-oriented database (OODB) can be seen as the special case of (b) above.

Therefore, a class is fuzzy because of the following several reasons. First, some objects are fuzzy ones, which have similar properties. A class defined by these objects may be fuzzy. These objects belong to the class with membership degree of [0, 1]. Second, when a class is intensionally defined, the domain of an attribute may be fuzzy and a fuzzy class is formed. For example, a class *Old equipment* is a fuzzy one because the domain of its attribute *Using period* is a set of fuzzy values such as *long*, *very long*, and *about* 20 *years*. Third, the subclass produced by a fuzzy class by means of specialization and the superclass produced by some classes (in which there is at least one class who is fuzzy) by means of generalization are also fuzzy.

The main difference between fuzzy classes and crisp classes is that the boundaries of fuzzy classes are imprecise. The imprecision in the class boundaries is caused by the imprecision of the values in the attribute domain. In the *FOODB*, classes are fuzzy because their attribute domains are fuzzy. The issue that an object fuzzily belongs to a class occurs since a class or an object is fuzzy. Similarly, a class is a subclass of another class with membership degree of [0, 1] because of the class fuzziness. In the *OODB*, the above-mentioned relationships are certain. Therefore, the evaluations of fuzzy object-class relationships and fuzzy inheritance hierarchies are the cores of information modeling in the *FOODB*.

2.5.3 Fuzzy Object-Class Relationships

In the *FOODB*, the following four situations can be distinguished for object-class relationships.

(a) Crisp class and crisp object. This situation is the same as the *OODB*, where the object belongs or not to the class certainly. For example, the objects *Car* and *Computer* are for a class *Vehicle*, respectively.

(b) Crisp class and fuzzy object. Although the class is precisely defined and has the precise boundary, an object is fuzzy since its attribute value(s) may be fuzzy. In this situation, the object may be related to the class with the special degree in [0, 1]. For example, the object which *position* attribute may be *graduate*, *research assistant*, or *research assistant professor*, is for the class *Faculty*.

(c) Fuzzy class and crisp object. Being the same as the case in (b), the object may belong to the class with the membership degree in [0, 1]. For example, a Ph.D. student is for *Young student* class.
(d) Fuzzy class and fuzzy object. In this situation, the object also belongs to the class with the membership degree in [0, 1].

The object-class relationships in (b), (c) and (d) above are called *fuzzy object-class relationships*. In fact, the situation in (a) can be seen the special case of fuzzy object-class relationships, where the membership degree of the object to the class is one. It is clear that estimating the membership of an object to the class is crucial for fuzzy object-class relationship when classes are instantiated.

In the *OODB*, determining if an object belongs to a class depends on if its attribute values are respectively included in the corresponding attribute domains of the class. Similarly, in order to calculate the membership degree of an object to the class in a fuzzy object-class relationship, it is necessary to evaluate the degrees that the attribute domains of the class include the attribute values of the object. However, it should be noted that in a fuzzy object-class relationship, only the inclusion degree of object values with respect to the class domains is not accurate for the evaluation of membership degree of an object to the class. The attributes play different role in the definition and identification of a class. Some may be dominant and some not. Therefore, a weight *w* is assigned to each attribute of the class according to its importance by designer. Then the membership degree of an object to the class in a fuzzy object-class relationship should be calculated using the inclusion degree of object values with respect to the class domains, and the weight of attributes.

Let *C* be a class with attributes $\{A_1, A_2, \ldots, A_n\}$, *o* be an object on attribute set $\{A_1, A_2, \ldots, A_n\}$, and $o\ (A_i)$ denote the attribute value of *o* on A_i $(1 \le i \le n)$. In *C*, each attribute A_i is connected with a domain denoted *dom* (A_i). The inclusion degree of $o\ (A_i)$ with respect to *dom* (A_i) is denoted ID (*dom* (A_i), $o\ (A_i)$). In the following, we investigate the evaluation of ID (*dom* (A_i), $o\ (A_i)$). As we know, *dom* (A_i) is a set of crisp values in the *OODB* and may be a set of fuzzy subsets in fuzzy databases. Therefore, in a uniform *OODB* for crisp and fuzzy information modeling, *dom* (A_i) should be the union of these two components, *dom* (A_i) = *cdom* $(A_i) \cup$ *fdom* (A_i), where *cdom* (A_i) and *fdom* (A_i) respectively denote the sets of crisp values and fuzzy subsets. On the other hand, $o\ (A_i)$ may be a crisp value or a fuzzy value. The following cases can be identified for evaluating ID (*dom* (A_i), $o\ (A_i)$).

Case 1 $o\ (A_i)$ is a fuzzy value. Let *fdom* $(A_i) = \{f_1, f_2, \ldots, f_m\}$, where f_j $(1 \le j \le m)$ is a fuzzy value, and *cdom* $(A_i) = \{c_1, c_2, \ldots, c_k\}$, where c_l $(1 \le l \le k)$ is a crisp value. Then

ID (*dom* (A_i), $o\ (A_i)$) = max (ID (*cdom* (A_i), $o\ (A_i)$), ID (*fdom* (A_i), $o\ (A_i)$)) = max (SID $(\{1.0/c_1, 1.0/c_2, \ldots, 1.0/c_k\}, o\ (A_i))$, $\max_j$ (SID $(f_j, o\ (A_i))$)),

where SID (x, y) is used to calculate the degree that fuzzy value *x* include fuzzy value *y*.

Case 2 o (A_i) is a crisp value. Then

ID (*dom* (A_i), o (A_i)) = 1 if o (A_i) ∈ *cdom* (A_i) else
ID (*dom* (A_i), o (A_i)) = ID (*fdom* (A_i), {1.0/o (A_i)}).

Consider a fuzzy class *Young students* with attributes *Age* and *Height*, and two objects o_1 and o_2. Assume *cdom* (*Age*) = {5 − 20}, *fdom* (*Age*) = {{1.0/20, 1.0/21, 0.7/22, 0.5/23}, {0.4/22, 0.6/23, 0.8/24, 1.0/25, 0.9/26, 0.8/27, 0.6/28}, {0.6/27, 0.8/28, 0.9/29, 1.0/30, 0.9/31, 0.6/32, 0.4/33, 0.2/34}}, and *dom* (*Height*) = *cdom* (*Height*) = [60, 210]. Let o_1 (*Age*) = 15, o_2 (*Age*) = {0.6/25, 0.8/26, 1.0/27, 0.9/28, 0.7/29, 0.5/30, 0.3/31}, and o_2 (*Height*) = 182. According to the definition above, we have

ID (*dom* (*Age*), o_1 (*Age*)) = 1,
ID (*dom* (*Height*), o_2 (*Height*)) = 1,
ID (*cdom* (*Age*), o_2 (*Age*)) = SID ({1.0/5, 1.0/6, …, 1.0/19, 1.0/20}, o_2 (*Age*)) = 0, and
ID (*fdom* (*Age*), o_2 (*Age*)) = max (SID ({1.0/20, 1.0/21, 0.7/22, 0.5/23}, o_2 (*Age*)), SID {0.4/22, 0.6/23, 0.8/24, 1.0/25, 0.9/26, 0.8/27, 0.6/28}, o_2 (*Age*)), SID ({0.6/27, 0.8/28, 0.9/29, 1.0/30, 0.9/31, 0.6/32, 0.4/33, 0.2/34}, o_2 (*Age*))) = max (0, 0.58, 0.60) = 0.60.

Therefore,

ID (*dom* (*Age*), o_2 (*Age*)) = max (ID (*cdom* (*Age*), o_2 (*Age*)), ID (*fdom* (*Age*), o_2 (*Age*))) = 0.60.

Now, we define the formula to calculate the membership degree of the object o to the class C as follows, where w (A_i (C)) denotes the weight of attribute A_i to class C.

$$\mu_C(o) = \frac{\sum_{i=1}^{n} ID(dom\ (A_i), o\ (A_i)) \times w\ (A_i(C))}{\sum_{i=1}^{n} w\ (A_i(C))}$$

Consider the fuzzy class *Young students* and object o_2 above. Assume w (*Age* (*Young students*)) = 0.9 and w (*Height* (*Young students*)) = 0.2. Then

$$\mu_{\text{Youngstudents}}(o_2) = (0.9 \times 0.6 + 0.2 \times 1.0)/(0.9 + 0.2) = 0.67.$$

In the above-given determination that an object belongs to a class fuzzily, it is assumed that the object and the class have the same attributes, namely, class C is with attributes {A_1, A_2, …, A_n} and object o is on {A_1, A_2, …, A_n} also. Such an object-class relationship is called *direct object-class relationship*. As we know, there exist subclass/superclass relationships in the *OODB*, where subclass inherits some attributes and methods of the superclass, overrides some attributes and methods of the superclass, and define some new attributes and methods. Any object belonging to the subclass must belong to the superclass since a subclass is the specialization of the superclass. So we have one kind of special object-class relationship: the relationship between superclass and the object of subclass. Such an object-class relationship is called *indirect object-class relationship.* Since the object and the class in indirect object-class relationship have different attributes, in the following, we present how to calculate the membership degree of an object to the class in an indirect object-class relationship.

Let C be a class with attributes $\{A_1, A_2, \ldots, A_k, A_{k+1}, \ldots, A_m\}$ and o be an object on attributes $\{A_1, A_2, \ldots, A_k, A'_{k+1}, \ldots, A'_m, A_{m+1}, \ldots, A_n\}$. Here attributes $A'_{k+1}, \ldots,$ and A'_m are overridden from $A_{k+1}, \ldots,$ and A_m and attributes $A_{m+1}, \ldots,$ and A_n are special. Then we have

$$\mu_C(o) = \frac{\sum_{i=1}^{k} ID(dom\ (A_i), o\ (A_i)) \times w\ (A_i(C)) + \sum_{j=k+1}^{m} ID(dom\ (A_j), o\ (A'_j)) \times w\ (A_j(C))}{\sum_{i=1}^{m} w\ (A_i(C))}.$$

Based on the direct object-class relationship and the indirect object-class relationship, now we focus on arbitrary object-class relationship. Let C be a class with attributes $\{A_1, A_2, \ldots, A_k, A_{k+1}, \ldots, A_m, A_{m+1}, \ldots, A_n\}$ and o be an object on attributes $\{A_1, A_2, \ldots, A_k, A'_{k+1}, \ldots, A'_m, B_{m+1}, \ldots, B_p\}$. Here attributes $A'_{k+1}, \ldots,$ and A'_m are overridden from $A_{k+1}, \ldots,$ and A_m, or $A_{k+1}, \ldots,$ and A_m are overridden from $A'_{k+1}, \ldots,$ and A'_m. Attributes $A_{m+1}, \ldots,$ and A_n and $B_{m+1}, \ldots, B_p$ are special in $\{A_1, A_2, \ldots, A_k, A_{k+1}, \ldots, A_m, A_{m+1}, \ldots, A_n\}$ and $\{A_1, A_2, \ldots, A_k, A'_{k+1}, \ldots, A'_m, B_{m+1}, \ldots, B_p\}$, respectively. Then we have

$$\mu_C(o) = \frac{\sum_{i=1}^{k} ID(dom\ (A_i), o\ (A_i)) \times w\ (A_i(C)) + \sum_{j=k+1}^{m} ID(dom\ (A_j), o\ (A'_j)) \times w\ (A_j(C))}{\sum_{i=1}^{n} w\ (A_i(C))}.$$

Since an object may belong to a class with membership degree in [0, 1] in fuzzy object-class relationship, it is possible that an object that is in a direct object-class relationship and an indirect object-class relationship simultaneously belongs to the subclass and superclass with different membership degrees. This situation occurs in fuzzy inheritance hierarchies, which will be investigated in next section. Also for two classes that do not have subclass/superclass relationship, it is possible that an object may belong to these two classes with different membership degrees simultaneously. This situation only arises in fuzzy object-oriented databases. In the *OODB*, an object may or may not belong to a given class definitely. If it belongs to a given class, it can only belong to it uniquely (except for the case of subclass/superclass).

The situation where an object belongs to different classes with different membership degrees simultaneously in fuzzy object-class relationships is called *multiple membership of object* in this chapter. Now let us focus on how to handle the multiple membership of object in fuzzy object-class relationships. Let C_1 and

C_2 be (fuzzy) classes and α be a given threshold. Assume there exists an object o. If $\mu_{C1}(o) \geq \alpha$ and $\mu_{C2}(o) \geq \alpha$, the conflict of the multiple membership of object occurs, namely, o belongs to multiple classes simultaneously. At this moment, which one in C_1 and C_2 is the class of object o dependents on the following cases.

Case 1 There exists a direct object-class relationship between object o and one class in C_1 and C_2.
Then the class in the direct object-class relationship is the class of object o.

Case 2 There is no direct object-class relationship but only an indirect object-class relationship between object o and one class in C_1 and C_2, say C_1. And there exists such subclass $C_1^{'}$ of C_1 that object o and $C_1^{'}$ are in a direct object-class relationship.
Then class $C_1^{'}$ is the class of object o.

Case 3 There is neither direct object-class relationship nor indirect object-class relationship between object o and classes C_1 and C_2. Or there exists only an indirect object-class relationship between object o and one class in C_1 and C_2, say C_1, but there is not such subclass $C_1^{'}$ of C_1 that object o and $C_1^{'}$ are in a direct object-class relationship.
Then class C_1 is considered as the class of object o if $\mu_{C1}(o) > \mu_{C2}(o)$, else class C_2 is considered as the class of object o.

It can be seen that in Case 1 and Case 2, the class in direct object-class relationship is always the class of object o and the object and the class have the same attributes. In Case 3, however, object o and the class that is considered as the class of object o, say C_1, have different attributes. It should be pointed out that class C_1 and object o are definitely defined, respectively, viewed from their structures. For the situation in Case 3, the attributes of C_1 do not affect the attributes of o and the attributes of o do not affect the attributes of C_1 also. There should be a class C and C and o are in direct object-class relationship. But class C is not available so far. That C_1 is considered as the class of object o, compared with C_2, only means that C_1 is more similar to C than C_2. Class C is the class of object o once C is available.

Consider three fuzzy classes C_1 with $\{A, B\}$, C_2 with $\{A, B, D\}$, and C_3 with $\{A, F\}$. There exists a fuzzy object o on $\{A, B', E\}$. Here, B' is overridden from B and $D \neq E \neq F$. According to the definitions above, we have

$$\mu_{C_1}(o) = \frac{ID(dom(A), o(A)) \times w(A(C_1)) + ID(dom(B), o(B')) \times w(B(C_1))}{w(A(C_1)) + w(B(C_1))},$$

$$\mu_{C_2}(o) = \frac{ID(dom(A), o(A)) \times w(A(C_2)) + ID(dom(B), o(B')) \times w(B(C_2))}{w(A(C_2)) + w(B(C_2)) + w(D(C_2))},$$

$$\mu_{C_3}(o) = \frac{ID(dom(A), o(A)) \times w(A(C_3))}{w(A(C_3)) + w(F(C_3))}.$$

Assume

$w\ (A\ (C_1)) = w\ (A\ (C_2)) = w\ (A\ (C_3))$,
$w\ (B\ (C_1)) = w\ (B\ (C_2))$, and
$w\ (B\ (C_2)) + w\ (D\ (C_2)) = w\ (F\ (C_3))$.

Also assume $\mu_{C1}\ (o) \geq \alpha$, $\mu_{C2}\ (o) \geq \alpha$, and $\mu_{C3}\ (o) \geq \alpha$, where α is a given threshold. Then object o belongs to classes C_1, C_2 and C_3 simultaneously. The conflict of the multiple membership of object occurs. It can be seen that the relationship between o and C_1 is an indirect object-class relationship. But the relationship between o and C_2, which is the subclass of class C_1, is not a direct object-class relationship. So class C_2 is not the class of object o. It can also be seen that $\mu_{C1}\ (o) \geq \mu_{C2}\ (o) \geq \mu_{C3}\ (o)$. So C_1 is considered as the class of object o. But in fact, there should be a new class C with $\{A, B', E\}$, which is the class in the direct object-class relationship of o and C. That $\mu_{C1}\ (o) \geq \mu_{C2}\ (o) \geq \mu_{C3}\ (o)$ only means that C_1 with $\{A, B\}$ is more similar to C with $\{A, B', E\}$ than C_2 with $\{A, B, E\}$ and C_3 with $\{A, F\}$. When class C is not available right now, class C_1 is considered as the class of object o.

2.5.4 Fuzzy Inheritance Hierarchies

In the *OODB*, a new class, called subclass, is produced from another class, called superclass by means of inheriting some attributes and methods of the superclass, overriding some attributes and methods of the superclass, and defining some new attributes and methods. Since a subclass is the specialization of the superclass, any one object belonging to the subclass must belong to the superclass. This characteristic can be used to determine if two classes have subclass/superclass relationship.

In the *FOODB*, however, classes may be fuzzy. A class produced from a fuzzy class must be fuzzy. If the former is still called subclass and the later superclass, the subclass/superclass relationship is fuzzy. In other words, a class is a subclass of another class with membership degree of [0, 1] at this moment. Correspondingly, the method used in the *OODB* for determination of subclass/superclass relationship is modified as

(a) for any (fuzzy) object, if the member degree that it belongs to the subclass is less than or equal to the member degree that it belongs to the superclass, and
(b) the member degree that it belongs to the subclass is great than or equal to the given threshold.

The subclass is then a subclass of the superclass with the membership degree, which is the minimum in the membership degrees to which these objects belong to the subclass.

Let C_1 and C_2 be (fuzzy) classes and β be a given threshold. We say C_2 is a subclass of C_1 if

$$(\forall o)\,(\beta \leq \mu_{C2}(o) \leq \mu_{C1}(o)).$$

The membership degree that C_2 is a subclass of C_1 should be $\min_{\mu C2(o) \geq \beta}(\mu_{C2}(o))$.

It can be seen that by utilizing the inclusion degree of objects to the class, we can assess fuzzy subclass/superclass relationships in the *FOODB*. It is clear that such assessment is indirect. If there is no any object available, this method is not used. In fact, the idea used in evaluating the membership degree of an object to a class can be used to determine the relationships between fuzzy subclass and superclass. We can calculate the inclusion degree of a (fuzzy) subclass with respect to the (fuzzy) superclass according to the inclusion degree of the attribute domains of the subclass with respect to the attribute domains of the superclass as well as the weight of attributes. In the following, we give the method for evaluating the inclusion degree of fuzzy attribute domains.

Let C_1 and C_2 be (fuzzy) classes with attributes $\{A_1, A_2, \ldots, A_k, A_{k+1}, \ldots, A_m\}$ and $\{A_1, A_2, \ldots, A_k, A'_{k+1}, \ldots, A'_m, A_{m+1}, \ldots, A_n\}$, respectively. It can be seen that in C_2, attributes $A_1, A_2, \ldots,$ and A_k are directly inherited from $A_1, A_2, \ldots,$ and A_k in C_1, attributes $A'_{k+1}, \ldots,$ and A'_m are overridden from $A_{k+1}, \ldots,$ and A_m in C_1, and attributes $A_{m+1}, \ldots,$ and A_n are special. For each attribute in C_1 or C_2, say A_i, there is a domain, denoted *dom* (A_i). As shown above, *dom* (A_i) should be *dom* (A_i) = *cdom* (A_i) ∪ *fdom* (A_i), where *cdom* (A_i) and *fdom* (A_i) denote the sets of crisp values and fuzzy subsets, respectively. Let A_i and A_j be attributes of C_1 and C_2, respectively. The inclusion degree of *dom* (A_j) with respect to *dom* (A_i) is denoted by ID (*dom* (A_i), *dom* (A_j)). Then we identify the following cases and investigate the evaluation of ID (*dom* (A_i), *dom* (A_j)):

(a) when $i \neq j$ and $1 \leq i, j \leq k$, ID (*dom* (A_i), *dom* (A_j)) = 0,
(b) when $i = j$ and $1 \leq i, j \leq k$, ID (*dom* (A_i), *dom* (A_j)) = 1, and
(c) when $i = j$ and $k + 1 \leq i, j \leq m$, ID (*dom* (A_i), *dom* (A_j)) = ID (*dom* (A_i), *dom* (A'_i)) = max (ID (*dom* (A_i), *cdom* (A'_i)), ID (*dom* (A_i), *fdom* (A'_i))).

Now we respectively define ID (*dom* (A_i), *cdom* (A'_i)) and ID (*dom* (A_i), *fdom* (A'_i)). Let *fdom* (A'_i) = $\{f_1, f_2, \ldots, f_m\}$, where f_j $(1 \leq j \leq m)$ is a fuzzy value, and *cdom* (A'_i) = $\{c_1, c_2, \ldots, c_k\}$, where c_l $(1 \leq l \leq k)$ is a crisp value. We can consider $\{c_1, c_2, \ldots, c_k\}$ as a special fuzzy value $\{1.0/c_1, 1.0/c_2, \ldots, 1.0/c_k\}$. Then we have the following:

ID (*dom* (A_i), *cdom* (A'_i)) = ID (*dom* (A_i), $\{1.0/c_1, 1.0/c_2, \ldots, 1.0/c_k\}$).
ID (*dom* (A_i), *fdom* (A'_i)) = $\max_j$ (ID (*dom* (A_i), f_j)).

Based on the inclusion degree of attribute domains of the subclass with respect to the attribute domains of its superclass as well as the weight of attributes, we can define the formula to calculate the degree to which a fuzzy class is a subclass of another fuzzy class. Let C_1 and C_2 be (fuzzy) classes with attributes $\{A_1, A_2, \ldots, A_k, A_{k+1}, \ldots, A_m\}$ and $\{A_1, A_2, \ldots, A_k, A'_{k+1}, \ldots, A'_m, A_{m+1}, \ldots, A_n\}$, respectively, and w (A) denote the weight of attribute A. Then the degree that C_2 is the subclass of C_1, written μ (C_1, C_2), is defined as follows.

$$\mu(C_1, C_2) = \frac{\sum_{i=1}^{m} ID(dom\ (A_i(C_1)), dom\ (A_i(C_2))) \times w\ (A_i)}{\sum_{i=1}^{m} w\ (A_i)}$$

In subclass-superclass hierarchies, a critical issue is multiple inheritance of class. Ambiguity arises when more than one of the superclasses have common attributes and the subclass does not declare explicitly the class from which the attribute was inherited.

Let class C be a subclass of classes C_1 and C_2. Assume that the attribute A_i in C_1, denoted by A_i (C_1), is common to the attribute A_i in C_2, denoted by A_i (C_2). If *dom* (A_i (C_1)) and *dom* (A_i (C_2)) are identical, there does not exist a conflict in the multiple inheritance hierarchy and C inherits attribute A_i directly. If *dom* (A_i (C_1)) and *dom* (A_i (C_2)) are not identical, however, the conflict occurs. At this moment, which one in A_i (C_1) and A_i (C_2) is inherited by C dependents on the following rule:

If ID (dom (A_i (C_1)), dom (A_i (C_2))) × w (A_i (C_1)) > ID (dom (A_i (C_2)), dom (A_i (C_1))) × w (A_i (C_2)), then A_i (C_1) is inherited by C, else A_i (C_2) is inherited by C.

Note that in fuzzy multiple inheritance hierarchy, the subclass has different degrees with respect to different superclasses, not being the same as the situation in classical object-oriented database systems.

2.5.5 Algebraic Operations in Fuzzy Object-Oriented Databases

In algebraic operations, a basic task is to determine the semantic relationship between two objects and assess if they are duplicate. Therefore, in the following we first introduce how to assess and deal with fuzzy object redundancies (Yan et al. 2012).

The notion of the semantic inclusion degree and equivalence degree of attribute data introduced in Sect. 2.4.2 can be extended to fuzzy objects. Let o_i = <a_{i1}, a_{i2}, …, a_{in}, μ_i> and o_j = <a_{j1}, a_{j2}, …, a_{jn}, μ_j> be two objects in fuzzy class C which contains attribute set $\{A_1, A_2, \ldots, A_n, \Lambda\}$. In class C, attribute name Λ is used to denote its membership attribute. As a result, μ_i in o_i is the membership degree that o_i belongs to C, and μ_j in o_j is the membership degree that o_j belongs to C. We use o_i [A_m] ($1 \leq m \leq n$) to denote the attribute value of object o_i on attribute A_m. It is clear that o_i [A_m] = a_{im} and o_i [$\varLambda$] = μ_i ($0 < m \leq 1$). The semantic inclusion degree of objects o_i and o_j is denoted

$$\begin{aligned}\mathrm{SID}\ (o_i, o_j) = {} & \mathrm{SID}\ (o_i[\mathrm{A_1}], o_j[\mathrm{A_1}]) \times \mathrm{w_1} + \mathrm{SID}\ (o_i[\mathrm{A_2}], o_j[\mathrm{A_2}]) \times \mathrm{w_2} + \ldots \\ & + \mathrm{SID}\ (o_i[\mathrm{A_n}], o_j[\mathrm{A_n}]) \times \mathrm{w}_n.\end{aligned}$$

The semantic equivalence degree of objects o_i and o_j is denoted

$$\begin{aligned}\mathrm{SE}\ (o_i, o_j) = {} & \mathrm{SE}\ (o_i[\mathrm{A_1}], o_j[\mathrm{A_1}]) \times \mathrm{w_1} + \mathrm{SE}\ (o_i[\mathrm{A_2}], o_j[\mathrm{A_2}]) \times \mathrm{w_2} + \ldots \\ & + \mathrm{SE}\ (o_i[\mathrm{A_n}], o_j[\mathrm{A_n}]) \times \mathrm{w}_n.\end{aligned}$$

It can be seen that, in order to calculate the inclusion degree and equivalence degree of these two objects, except for the membership attribute, we need to compare each pair of other attribute values of two objects on the same attributes.

As we know, in the object-oriented databases, the attribute values of objects may be complex values and even be other objects. For such a pair of values on an attribute, say A_m ($1 \leq m \leq n$), we cannot calculate SID (o_i [A_m], o_j [A_m]) by applying the definition of SID (π_A, π_B) above. But complex values are eventually consisted of some simple values. For these simple values, we can directly calculate their inclusion degrees and then can obtain the inclusion degree of complex values by combining the inclusion degrees of these simple values. Suppose that o_i [A_m] and o_j [A_m] are fuzzy complex values, in which o_i [A_m] = <a_{im1}, a_{im2}, …, a_{imk}>, o_j [A_m] = <a_{jm1}, a_{jm2}, …, a_{jmk}>, and a_{iml} and a_{jml} ($1 \leq l \leq k$) are simple values. Then we can directly calculate SID (a_{iml}, a_{jml}) and finally have

$$\mathrm{SID}\ (o_i[\mathrm{A_m}], o_j[\mathrm{A_m}]) = \min(\mathrm{SID}\ (\mathrm{a_{im1}},\ \mathrm{a_{jm1}}),\ \mathrm{SID}\ (\mathrm{a_{im2}},\ \mathrm{a_{jm2}}), \ldots,\ \mathrm{SID}\ (\mathrm{a_{imk}},\ \mathrm{a_{jmk}}))$$

Based on the inclusion degree and equivalence degree of two fuzzy objects, we can deal with fuzzy object redundancy. The processing of fuzzy value redundancy can also be extended to that of fuzzy object redundancy. In a similar way, fuzzy object redundancy can be classified into *inclusion redundancy* and *equivalence redundancy* of fuzzy objects.

Let C be a fuzzy class with attribute set {A_1, A_2, …, A_n, $\varLambda$}. Let o_i = <a_{i1}, a_{i2}, …, a_{in}, μ_i> and o_j = <a_{j1}, a_{j2}, …, a_{jn}, μ_j> be two objects in C. Then with a given threshold $\beta \in [0, 1]$, object o_j is inclusively β-redundant to object o_i if and only if SID (o_i, o_j) $\geq \beta$ holds true, and objects o_i and o_j are equivalently β-redundant if and only if SE (o_i, o_j) $\geq \beta$ holds true.

When o_i and o_j are inclusively redundant or equivalently redundant, the removal of redundancy can be achieved by merging o_i and o_j and producing a new fuzzy class o. Being similar to the merging of fuzzy data above, we have three kinds of merging operations for fuzzy objects in order to meet different requirements of object manipulations.

$$o = merge_{\cup}(o_i, o_j) = <merge_{\cup}(o_i[A_1], o_j[A_1]), merge_{\cup}(o_i[A_2], o_j[A_2]), \ldots, merge_{\cup}(o_i[A_n], o_j[A_n]), \max(o_i[\Lambda], o_j[\Lambda]) >$$

$$o = merge_{-}(o_i, o_j) = <merge_{-}(o_i[A_1], o_j[A_1]), merge_{-}(o_i[A_2], o_j[A_2]), \ldots, merge_{-}(o_i[A_n], o_j[A_n]), \max(o_i[\Lambda] - o_j[\Lambda], 0) >$$

$$o = merge_{\cap}(o_i, o_j) = <merge_{\cap}(o_i[A_1], o_j[A_1]), merge_{\cap}(o_i[A_2], o_j[A_2]), \ldots, merge_{\cap}(o_i[A_n], o_j[A_n]), \min(o_i[\Lambda], o_j[\Lambda]) >$$

Based on the notion of the fuzzy object redundancies above, next we describe the fuzzy algebraic operations (Yan et al. 2012). After referring to the fuzzy algebraic operations in the fuzzy relational databases (Umano and Fukami 1994; Ma and Mili 2002), we classify the fuzzy algebraic operations of the fuzzy object-oriented databases into two types: algebraic operations for fuzzy classes and algebraic operations for fuzzy objects.

In order to define the algebraic operations for fuzzy objects and fuzzy classes, we first introduce some notations being used below. Let *C* be a (fuzzy) class and *Attr* (*C*) be its attribute set. Then we have *Attr′* (*C*) which is obtained from *Attr* (*C*) through removing the membership degree attributes from *Attr* (*C*). We use Λ_C to represent the membership degree attribute of *C*. It is clear that *Attr′* (*C*) = *Attr* (*C*) − {Λ_C}. So *Attr′* (*C*) actually contains all general attribute of class *C*, say {A_1, A_2, …, A_n}, not including its membership degree attribute Λ_C. Class *C* contains a set of (fuzzy) objects, denoted $C = \{o_1, o_2, \ldots, o_m\}$. We use *o* (*C*) to present an object *o* of *C*. For a general attribute in *C*, say A_i ($1 \leq i \leq n$), *o* [A_i] is used to represent the value of *o* on A_i. The value of *o* on Λ_C is *o* [Λ_C], which is the membership degree that *o* belongs to *C*. Generally μ_C (*o*) is used to represent the membership degree that *o* belongs to *C*.

2.5.5.1 Algebraic Operation for Fuzzy Objects

The algebraic operation for fuzzy objects is eventual the *fuzzy selection* (σ). A selection operation refers to such a procedure that the objects of the classes satisfying a given selection condition are selected. The major issue here is how to determine if an object satisfies the selection condition. We start with the syntax of selection condition for fuzzy object selection.

In the classical relational databases, a predicate *P* denoted the selection condition is formed through combining the basic clause "X θ Y" as operands with operators $\neg$, $\wedge$, and $\vee$, in which $\theta \in \{>, <, =, \neq, \geq, \leq\}$, X is an attribute, and Y may be a constant, an attributes or an expression which is formed through combining constants, attributes or expressions with arithmetic operations. For the fuzzy databases, in "X θ Y", the attribute and constant may be fuzzy, and also the "θ" may be fuzzy comparison operations such as $\succ_\beta$, $\prec_\beta$, $\succcurlyeq_\beta$, $\preccurlyeq_\beta$, $\approx_\beta$, and $\not\approx_\beta$, where β is a given threshold. A basic fuzzy expression is hereby formed and used to further form a fuzzy predicate, denoted P_f, as a fuzzy selection condition.

To evaluate the fuzzy expressions, it is necessary to discuss the semantics of fuzzy "X θ Y". Let π_A and π_B be two fuzzy data over $U = \{u_1, u_2, \ldots, u_n\}$. Then

(a) $\pi_A \approx_\beta \pi_B$ if SE $(\pi_A, \pi_B) \geq \beta$,
(b) $\pi_A \not\approx_\beta \pi_B$ if SE $(\pi_A, \pi_B) < \beta$,
(c) $\pi_A \succ_\beta \pi_B$ if $\pi_A \not\approx_\beta \pi_B$ and max (supp (π_A)) > max (supp (π_B)),
(d) $\pi_A \succcurlyeq_\beta \pi_B$ if $\pi_A \approx_\beta \pi_B$ or $\pi_A \succ_\beta \pi_B$,
(e) $\pi_A \prec_\beta \pi_B$ if $\pi_A \not\approx_\beta \pi_B$ and max (supp (π_A)) < max (supp (π_B)), and
(f) $\pi_A \preccurlyeq_\beta \pi_B$ if $\pi_A \not\approx_\beta \pi_B$ or $\pi_A \prec_\beta \pi_B$.

Here "supp (π_A)" is used to define the support set of the possibility distribution π_A. Let π_A be a fuzzy data on U based on possibility distribution and $\pi_A(u)$, $u \in U$, denote the possibility degree that u is true. The set of the elements that have non-zero possibility degrees in π_A is called the support of π_A. Formally

$$\text{supp}\ (\pi_A) = \{u \mid u \in U \text{ and } \pi_A(u) > 0\}$$

The fuzzy comparison operations $\succ_\beta$, $\prec_\beta$, $\succcurlyeq_\beta$, $\preccurlyeq_\beta$, $\approx_\beta$, and $\not\approx_\beta$ have the same properties as the classical comparison operations >, <, $\geq$, $\leq$, =, and $\neq$. For example, $\pi_A \preccurlyeq_\beta \pi_B$ is equivalent to $(\pi_A \prec_\beta \pi_B) \vee (\pi_A \approx_\beta \pi_B)$ and to $\neg\ (\pi_A \succ_\beta \pi_B)$.

Let C be a fuzzy class and P_f be a fuzzy predicate denoted the selection condition. Then the selection on C for P_f is defined as follows.

$$\sigma_{Pf}(C) = \{o(C) \mid o(C) \wedge P_f(o)\}$$

For example, suppose that we have a fuzzy class C', in which $Attr'$ (C') = {ID, Department, Age, Degree, Nationality, Office}, and

o'_1 (C') = <ID: 9106, Department: CS, Age: {(18, 0.5), (19, 0.9), (20, 0.7)}, Degree: M. Phil, Nationality: USA, Office: Y1101, Λ_{C1}: 0.8 >,
o'_2 (C') = <ID: 9107, Department: CS, Age: {(20, 0.4), (21, 0.8), (22, 0.6)}, Degree: M. Phil, Nationality: Canada, Office: Y1101, Λ_{C1}: 0.9 >,
o'_3 (C') = <ID: 9705, Department: IS, Age: {(24, 0.2), (25, 0.9), (26, 1.0)}, Degree: M. Phil, Nationality: France, Office: B6280, Λ_{C1}: 0.8 >,
o'_4 (C') = <ID: 9706, Department: IS, Age: {(28, 0.2), (29, 0.7), (30, 0.6)}, Degree: Ph.D., Nationality: Italy, Office: B6280, Λ_{C1}: 0.7 >, and
o'_5 (C') = <ID: 9707, Department: IS, Age: {(29, 0.6), (30, 0.7), (31, 0.2)}, Degree: Ph.D., Nationality: France, Office: B6280, Λ_{C1}: 0.6 >.

Now for a given threshold $\beta = 0.8$ and a given selection condition P_f: *Department* = *"IS"* $\wedge$ *Age* $\approx$ *"{1.0/25, 0.9/26, 0.3/27}"*, we can draw that o'_3 (C') satisfies the selection condition and is returned, in which

SE (o'_3 [Department], "IS") = 1 $\geq \beta$ and SE (o_3 [Age], {1.0/25, 0.9/26, 0.3/27}) = 0.82 $\geq \beta$.

2.5.5.2 Algebraic Operation for Fuzzy Classes

We can identify two kinds of algebraic operations for fuzzy classes: one is for single class and another is for multiple classes. The latter actually contains several combination operations of the fuzzy classes, which combine several (fuzzy) classes into a new class. Depending on the relationships between the attribute sets of the combining classes, five kinds of combination operations can be identified: *fuzzy product* (×), *fuzzy join* (⋈), *fuzzy union* (∪), *fuzzy difference* (−) and *fuzzy Intersection* (∩). The algebraic operation for single fuzzy class is eventual the *fuzzy projection* (Π). In the following, we give the formal definitions of fuzzy product, fuzzy join, fuzzy union, fuzzy intersection, fuzzy difference and fuzzy projection operations.

Fuzzy product The fuzzy product of C_1 and C_2 is a new class C, which is composed of these general attributes in C_1 and C_2 as well as a membership attribute. Generally it is required that $Attr'\ (C_1) \cap Attr'\ (C_2) = \Phi$ in the fuzzy product. The objects of C are created by the composition of objects from C_1 and C_2, in which C contains attributes $Attr'\ (C_1)$ and $Attr'\ (C_2)$ as well as Λ_C.

$$\begin{aligned} C = C_1 \times C_2 = \{&o(C) \mid (\forall o_1)\,(\forall o_2)\,(o_1(C_1) \wedge o_2(C_2) \wedge o[Attr'\ (C_1)] \\ &= o_1[Attr'\ (C_1)] \wedge o[Attr'\ (C_2)] = o_2[Attr'\ (C_2)] \wedge o[\Lambda_C] \\ &= op(o_1[\Lambda_{C1}], o_2\ [\Lambda_{C2}]))\} \end{aligned}$$

Here, $op\ (o_1\ [\Lambda_{C1}], o_2\ [\Lambda_{C2}])$ may be defined as min $(o_1\ [\Lambda_{C1}], o_2\ [\Lambda_{C2}])$ or may be defined as $o_1\ [\Lambda_{C1}] \times o_2\ [\Lambda_{C2}]$.

Fuzzy join The join operation is to join two classes together by combining every pair of objects in two classes, which satisfy the given conditions. For two fuzzy classes C_1 and C_2 and a fuzzy conditional predicate P_f, their join $C_1 \bowtie_{Pf} C_2$ forms a new class C. Here P_f is a fuzzy conditional predicate in the form of "A θ B", where $\theta \in \{\succ_\beta, \prec_\beta, \succcurlyeq_\beta, \preccurlyeq_\beta, \approx_\beta, \not\approx_\beta\}$, A $\in$ $Attr'\ (C_1)$, and B $\in$ $Attr'\ (C_2)$. According to P_f as well as the relationship between $Attr'\ (C_1)$ and $Attr'\ (C_2)$, we can identify several different fuzzy join operation.

(a) $Attr'\ (C_1) \cap Attr'\ (C_2) = \Phi$: it means that there is not common attribute in C_1 and C_2.
At this point, the fuzzy join operation is actually a kind of conditional product: only the objects which are respectively from C_1 and C_2 and satisfy P_f are combined together. The new class C contains attributes $Attr'\ (C_1)$ and $Attr'\ (C_2)$ as well as Λ_C. The fuzzy join operation is defined as follows.

$$\begin{aligned} C &= C_1 \bowtie_{Pf} C_2 \\ &= \{o(C) \mid (\exists o_1)\,(\exists o_2)\,(o_1(C_1) \wedge o_2(C_2) \wedge P_f(o_1, o_2) \wedge o[Attr'\ (C_1)] \\ &= o_1[Attr'\ (C_1)] \wedge o[Attr'\ (C_2)] \\ &= o_2[Attr'\ (C_2)] \wedge o[\Lambda_C] = op(o_1[\Lambda_{C1}], o_2[\Lambda_{C2}]))\} \end{aligned}$$

Actually $C = C_1 \bowtie_{Pf} C_2 = \sigma_{Pf}\,(C_1 \times C_2)$.

(b) $Attr'(C_1) \cap Attr'(C_2) \neq \Phi$: it means that there are some common attribute in C_1 and C_2.

At this point, we can further identify two cases:

(i) In P_f, A and B do not belong to $Attr'(C_1) \cap Attr'(C_2)$ simultaneously and $Attr'(C_1) \neq Attr'(C_2)$. Then $Attr(C) = Attr'(C_1) \cup (Attr'(C_2) - (Attr'(C_1) \cap Attr'(C_2))) \cup \{\Lambda_C\}$ and we have

$$\begin{aligned} C &= C_1 \bowtie_{Pf} C_2 \\ &= \{o(C) \mid (\exists o_1)(\exists o_2)(o_1(C_1) \wedge o_2(C_2) \wedge P_f(o_1, o_2) \wedge o[Attr'(C_1) \\ &\quad - (Attr'(C_1) \cap Attr'(C_2))] = o_1[Attr'(C_1) - (Attr'(C_1) \cap Attr'(C_2))] \\ &\quad \wedge o[Attr'(C_1) \cap Attr'(C_2)] \\ &= merge_{\cap}(o_1[Attr'(C_1) \cap Attr'(C_2)], o_2[Attr'(C_1) \cap Attr'(C_2)]) \\ &\quad \wedge o[Attr'(C_2) - (Attr'(C_1) \cap Attr'(C_2))] \\ &= o_2[Attr'(C_2) - (Attr'(C_1) \cap Attr'(C_2))] \wedge o[\Lambda_C] \\ &= op(o_1[\Lambda_{C1}], o_2[\Lambda_{C2}]))\} \end{aligned}$$

(ii) In P_f, A and B belong to $Attr'(C_1) \cap Attr'(C_2)$ simultaneously and A θ B is A $\approx_\beta$ B. Then the fuzzy join becomes the fuzzy natural join (i.e., the fuzzy equi-join), denoted $C = C_1 \bowtie C_2$, and $Attr(C) = Attr'(C_1) \cup (Attr'(C_2) - (Attr'(C_1) \cap Attr'(C_2))) \cup \{\Lambda_C\}$. The objects of C are created by the composition of objects from C_1 and C_2, which are semantically equivalent on $Attr'(C_1) \cap Attr'(C_2)$ under the given thresholds. It should be noted that, however, $Attr'(C_1) \cap Attr'(C_2) \neq \Phi$ implies C_1 and C_2 have the same weights of attributes for the attributes in $Attr'(C_1) \cap Attr'(C_2)$. This is an additional requirement to be met in the case of the fuzzy join operation. Let β be the given threshold. Then we have

$$\begin{aligned} C &= C_1 \bowtie_\beta C_2 \\ &= \{o \mid (\exists o_1)(\exists o_2)(o_1(C_1) \wedge o_2(C_2) \wedge \mathrm{SE}\,(o_1[Attr'(C_1) \\ &\quad \cap Attr'(C_2)], o_2[Attr'(C_1) \cap Attr'(C_2)]) \geq \beta \wedge o[Attr'(C_1) - (Attr'(C_1) \\ &\quad \cap Attr'(C_2))] \\ &= o_1[Attr'(C_1) - (Attr'(C_1) \cap Attr'(C_2))][Attr'(C_1) \cap Attr'(C_2)] \\ &= merge_{\cap}(o_1[Attr'(C_1) \cap Attr'(C_2)], o_2[Attr'(C_1) \cap Attr'(C_2)])[Attr'(C_2) \\ &\quad - (Attr'(C_1) \cap Attr'(C_2))] \\ &= o_2[Attr'(C_2) - (Attr'(C_1) \cap Attr'(C_2))] \wedge o[\Lambda_C] \\ &= op(o_1[\Lambda_{C1}], o_2[\Lambda_{C2}]))\} \end{aligned}$$

For example, suppose that we have two fuzzy classes C_1 and C_2, in which $Attr'(C_1)$ = {ID, Department, Age} and $Attr'(C_2)$ = {Name, Degree, Age}. For C_1, we have

o_{11} (C_1) = <ID: 9107, Department: ME, Age: {(21, 0.8), (22, 0.7)}, Λ_{C1}: 0.7> and

o_{12} (C_1) = <ID: 9712, Department: ME, Age: {(32, 0.8), (33, 0.9), (34, 0.6)}, Λ_{C1}: 0.8>.

For C_2, we have

o_{21} (C_2) = <Name: Tom, Degree: Ph.D., Age: {(21, 1.0), (22, 0.7)}, Λ_{C2}: 0.6> and

o_{22} (C_2) = <Name: Jack, Degree: Ph.D., Age: {(32, 0.7), (33, 1.0), (34, 0.7)}, Λ_{C2}: 0.7>.

For a given threshold $\beta = 0.8$ and $C_1 \bowtie_\beta C_2$, we have *Attr* (C) = {ID, Name, Department, Degree, Age, Λ} and

o_1 (C) = <ID: 9107, Name: Tom, Department: ME, Degree: Ph.D., Age: {(21, 0.8), (22, 0.7)}, Λ: 0.6> and

o_2 (C) = <ID: 9712, Name: Jack, Department: ME, Degree: Ph.D., Age: {(32, 0.7), (33, 0.9), (34, 0.6)}, Λ: 0.7>.

Fuzzy union The fuzzy union of C_1 and C_2 requires $Attr'$ $(C_1) = Attr'$ (C_2), which implies that all corresponding attributes in C_1 and C_2 have the same weights. Let a new class C be the fuzzy union of C_1 and C_2. Then the objects of C are composed of three kinds of objects:

(a) The first two kinds of objects are such objects that directly come from one component class (e.g., C_1) and are not redundant with any object in another component class (e.g., C_2) under the given threshold.
(b) The last kind of objects is such objects that are the results of merging the redundant objects from two component classes under the given threshold.

Let β be the given threshold. Then we have

$$C = C_1 \cup_\beta C_2 = \{o(C) \mid (\forall o_2)(\exists o_1)(o_2(C_2) \wedge o_1(C_1) \wedge \text{SE}\ (o_1, o_2) < \beta \wedge o = o_1) \vee (\forall o_1)(\exists o_2)(o_1(C_1) \wedge o_2(C_2) \wedge \text{SE}\ (o_2, o_1) < \beta \wedge o = o_2) \vee (\exists o_1)(\exists o_2)(o_1(C_1) \wedge o_2(C_2) \wedge \text{SE}\ (o_1, o_2) \geq \beta \wedge o = merge_\cup(o_1, o_2))\}$$

For example, suppose that we have two fuzzy classes C_1 and C_2, in which $Attr'$ $(C_1) = Attr'$ (C_2) = {ID, Department, Age}. For C_1, we have

o_{11} (C_1) = <ID: 9106, Department: CS, Age: {(19, 0.3), (20, 0.8), (21, 0.7)}, Λ_{C1}: 0.7>,

o_{12} (C_1) = <ID: 9107, Department: CS, Age: {(30, 0.6), (31, 0.9), (32, 0.7)}, Λ_{C1}: 0.7>, and

o_{13} (C_1) = <ID: 9711, Department: ME, Age: {(32, 0.5), (33, 0.7), (34, 0.6)}, Λ_{C1}: 0.9>.

For C_2, we have

o_{21} (C_2) = <ID: 9106, Department: CS, Age: {(20, 0.8), (21, 0.7)}, Λ_{C2}: 0.6>,
o_{22} (C_2) = <ID: 9108, Department: CS, Age: {(32, 0.5), (33, 0.8), (34, 0.6)}, Λ_{C2}: 0.8>, and
o_{23} (C_2) = <ID: 9711, Department: ME, Age: {(32, 0.8), (33, 0.6), (34, 0.7)}, Λ_{C2}: 0.8>.

Then for a given threshold $\beta = 0.8$ and $C = C_1 \cup_\beta C_2$, we have *Attr* (C) = {ID, Department, Age, Λ} and

o_1 (C) = <ID: 9106, Department: CS, Age: {(19, 0.3), (20, 0.8), (21, 0.7)}, Λ: 0.7>,
o_2 (C) = <ID: 9107, Department: CS, Age: {(30, 0.6), (31, 0.9), (32, 0.7)}, Λ: 0.7>,
o_3 (C) = <ID: 9108, Department: CS, Age: {(32, 0.5), (33, 0.8), (34, 0.6)}, Λ: 0.8>, and
o_4 (C) = <ID: 9711, Department: ME, Age: {(32, 0.8), (33, 0.7), (34, 0.7)}, Λ: 0.9>.

Fuzzy difference The fuzzy difference of C_1 and C_2 also requires *Attr*′ (C_1) = *Attr*′ (C_2) and all corresponding attributes in C_1 and C_2 have the same weights. Let a new class C be the fuzzy difference of C_1 and C_2, and let β be the given threshold. Then we have

$$C = C_1 -_\beta C_2 = \{o(C) | (\forall o_2)(\exists o_1)(o_2(C_2) \wedge o_1(C_1) \wedge \text{SE}(o_1, o_2) < \beta \wedge o = o_1) \vee (\exists o_1)(\exists o_2)(o_1(C_1) \wedge o_2(C_2) \wedge \text{SE}(o_1, o_2) \geq \beta \wedge o = merge_-(o_1, o_2))\}$$

For example, suppose that we have the same C_1 and C_2 shown in the example above. Then for a given threshold $\beta = 0.8$ and $C = C_1 -_\beta C_2$, we have *Attr* (C) = {ID, Department, Age, Λ} and

o_1 (C) = <ID: 9106, Department: CS, Age: {(19, 0.3)}, Λ: 0.1>,
o_2 (C) = <ID: 9107, Department: CS, Age: {(30, 0.6), (31, 0.9), (32, 0.7)}, Λ: 0.7>, and
o_3 (C) = <ID: 9711, Department: ME, Age: {(33, 0.1)}, Λ: 0.1>.

Fuzzy intersection The fuzzy intersection of C_1 and C_2 is to combine the common objects of these two classes, which requires *Attr*′ (C_1) = *Attr*′ (C_2) and all corresponding attributes in C_1 and C_2 have the same weights also. Let a new class C be the fuzzy intersection of C_1 and C_2, and let β be the given threshold. Then we have

$$C = C_1 \cap_\beta C_2 = \{o(C) | (\exists o_1)(\exists o_2)(o_1(C_1) \wedge o_2(C_2) \wedge \text{SE}(o_1, o_2) \geq \beta \wedge o = merge_\cap(o_1, o_2))\}$$

For example, suppose that we have the same C_1 and C_2 shown in the example above. Then for a given threshold $\beta = 0.8$ and $C = C_1 \cap_\beta C_2$, we have *Attr* (C) = {ID, Department, Age, Λ} and

o_1 (*C*)= <ID: 9106, Department: CS, Age: {(20, 0.8), (21, 0.7)}, *Λ*: 0.6> and
o_2 (*C*) = <ID: 9711, Department: ME, Age: {(32, 0.5), (33, 0.6), (34, 0.6)}, *Λ*: 0.8>.

Fuzzy projection For a given class (say *C*′) and a given subset of attributes (say *S*) the class, the projection of *C*′ on *S* is to remove attributes *Attr* (*C*′) − *S* from *C*′ and only remain attributes *S* in *C*′, forming a new class *C*. It is clear that $S \subset Attr$ (*C*′) and *Attr* (*C*) = *S*. Since only attributes *S* in *C*′ are remained, each object in *C*′ becomes a new object without attribute values on *Attr* (*C*′) − *S*. There may be redundancies among these new objects. After removing possible redundancies, the new objects constitute class *C*. The projection of *C*′ on *S* is defined as follows.

$$C = \Pi_S(C') = \{o(C) | (\forall o')\, (o'(C') \wedge o[S] = o'[S] \wedge o = merge_{\cup}(o[S]))\}$$

For example, suppose that we have the same *C*′ shown in the example in Sect. 2.5.5.1. Then for a given threshold $\beta = 0.8$ and $C = \Pi_{\{\text{Dept, Age}\}}$ (*C*′), we have *Attr* (*C*) = {Department, Age} and

o_1 (*C*) = <Department: CS, Age: {(18, 0.5), (19, 0.9), (20, 0.7)}>,
o_2 (*C*) = <Department: CS, Age: {(20, 0.4), (21, 0.8), (22, 0.6)}>,
o_3 (*C*) = <Department: IS, Age: {(24, 0.2), (25, 0.9), (26, 1.0)}>, and
o_4 (*C*) = <Department: IS, Age: {(28, 0.2), (29, 0.7), (30, 0.6), (31, 0.2)}>.

Note that there exists object redundancy after projecting o'_4 (*C*′) and o'_5 (*C*′) of *C*′ on attributes {Department, Age}. The result of redundancy removal forms o_4 of *C*.

2.5.5.3 Other Algebraic Operation

In addition to the major algebraic operations introduced above, there are also several algebraic operations are useful for fuzzy object and class operations. In the following, we discuss two operations: renaming operation and outerunion operation.

Renaming This operation is used to change the names of some attributes of a class. Let *C* be a fuzzy class, and A and B be two attributes satisfying A ∈ *Attr* (*C*) and B ∉ *Attr* (*C*), where A and B have the same domain. Now we rename A into B and form a new class *C*′. It is clear that *Attr* (*C*′) = *Attr* (*C*) − {A} ∪ {B}. Then class *C* with A renamed to B is defined as

$$C' = \rho_{A \leftarrow B}(C) = \{o'(C') \mid (\forall o)\, (o(C) \wedge o'\, [Attr(C') - B] = o[Attr(C) - A] \wedge o'\,[B] = o[A])\}.$$

It means that when attribute A in class *C* is renamed into B, the objects of *C* are not changed at all. These objects constitute class *C*′ which is formed after renaming A in *C* into B.

The renaming operation above is defined for renaming a single attribute. The operation for renaming several attributes (i.e., a set of attributes) follows the same processing.

Outerunion The general union operation requires two classes union-compatible. That is two classes have the same attributes. But sometimes it is needed to combine two classes with different attributes with union operations. For example, we integrate heterogeneous multiple classes. At this point, we need a kind of outerunion operation.

Let C_1 and C_2 be two fuzzy classes and let $Attr'\ (C_1) \neq Attr'\ (C_2)$. Let $Z = Attr'\ (C_1) \cap Attr'\ (C_2)$, $X = Attr'\ (C_1) - Attr'\ (C_2)$ and $Y = Attr'\ (C_2) - Attr'\ (C_1)$. It is possible that $X = \Phi$ or $Y = \Phi$ but it is not possible that $X = \Phi$ and $Y = \Phi$, or $Attr'\ (C_1) = Attr'\ (C_2)$ and the operation turns to the general union operation. Without loss of generality, it is assumed that $X \neq \Phi$ and $Y \neq \Phi$. Then the outerunion of C_1 and C_2, denoted by $C_1 \tilde{\cup} C_2$, forms a new class C with *Attr* $(C) = ZXY\Lambda$. Let β be the given threshold and then we have

$$
\begin{aligned}
C = C_1 \tilde{\cup}_b C_2 = \{o(C) | & (\forall o_2)\ (\exists o_1)\ (o_2(C_2) \wedge o_1(C_1) \wedge \text{SE}\ (o_1[Z], o_2[Z])\\
& < \beta [Z] = o_1[Z] \wedge o[X] = o_1[X] \wedge o[Y] = \varphi [\Lambda]\\
& = o_1[\Lambda_{C1}]) \vee (\forall o_1)\ (\exists o_2)\ (o_1(C_1) \wedge o_2(C_2) \wedge \text{SE}(o_1[Z], o_2[Z])\\
& < \beta \wedge o[Z] = o_2[Z] \wedge o[X] = \varphi \wedge o[Y] = o_2[Y] \wedge o[\Lambda]\\
& = o_2[\Lambda_{C2}]) \vee (\exists o_1)\ (\exists o_2)\ (o_1(C_1) \wedge o_2(C_2) \wedge \text{SE}\ (o_1[Z], o_2[Z])\\
& \geq \beta \wedge o[Z] = merge_{\cup}(o_1[Z], o_2[Z]) \wedge o[X] = o_1[X] \wedge o[Y]\\
& = o_2[Y] \wedge o[\Lambda] = \max(o_1[\Lambda_{C1}], o_2[\Lambda_{C2}]))\}
\end{aligned}
$$

Being similar to the general union operation, the objects of the above C are composed of three kinds of objects:

(a) The first kind of objects originally comes from C_1, which are not redundant with any object of C_2 on attributes Z. These objects have null values, which are expressed by φ, on attributes Y in C.
(b) The second kind of objects originally comes from C_2, which are not redundant with any object of C_1 on attributes Z. These objects have null values, which are expressed by φ, on attributes X in C.
(c) The last kind of objects originally comes from C_1 and C_2 simultaneously, which are redundant each other on attributes Z. On attributes Z in C, these objects are combined using the merging operation under the given threshold.

As mentioned above, the outerunion operation turns to the general union operation when two classes have the same attributes. The outerunion operation turns to the product operation when two classes have completely different attributes, i.e., the intersection of the attribute sets of two classes is empty.

For example, suppose that we have two fuzzy classes C_1 and C_2, in which *Attr* $(C_1) = \{\text{ID, Office, Age}, \Lambda_{C1}\}$ and *Attr* $(C_2) = \{\text{ID, Degree, Age}, \Lambda_{C2}\}$. For C_1, we have

o_{11} (C_1) = <ID: 9106, Office: Y1415, Age: {(19, 0.3), (20, 0.8), (21, 0.7)}, Λ_{C1}: 0.6> and

o_{12} (C_1) = <ID: 9107, Office: B6280, Age: {(30, 0.6), (31, 0.9), (32, 0.7)}, Λ_{C1}: 0.8>.

For C_2, we have

o_{21} (C_2) = <ID: 9106, Degree: {(BE, 0.7), (MPh, 0.5)}, Age: {(19, 0.3), (20, 0.8), (21, 0.7)}, Λ_{C2}: 0.7> and

o_{22} (C_2) = <ID: 9108, Degree: {(MPh, 0.6), (Ph.D., 0.8)}, Age: {(32, 0.5), (33, 0.9), (34, 0.7)}, Λ_{C2}: 0.9>.

Then for a given threshold $\beta = 0.9$ and $C = C_1 \tilde{\cup}_\beta C_2$, we have *Attr* ($C$) = {ID, Office, Degree, Age, Λ}, and

o_1 (C) = <ID: 9106, Office: Y1415, Degree: {(BE, 0.7), (MPh, 0.5)}, Age: {(19, 0.3), (20, 0.8), (21, 0.7)}, Λ: 0.7>,

o_2 (C) = <ID: 9107, Office: B6280, Degree: φ, Age: {(30, 0.6), (31, 0.9), (32, 0.7)}, Λ: 0.8>, and

o_3 (C) = <ID: 9108, Office: φ, Degree: {(MPh, 0.6), (Ph.D., 0.8)}, Age: {(32, 0.5), (33, 0.9), (34, 0.7)}, Λ: 0.9>.

2.5.5.4 SQL-like Fuzzy Querying

Query processing in the object-oriented databases refers to such a procedure that the objects satisfying a given condition are selected and then they are delivered to the user according to the required formats. These format requirements include which attributes appear in the result and if the result is grouped and ordered over the given attribute(s). So a query can be seen as comprising two components, namely, a boolean query condition and some format requirements. For the sake of the simple illustration, some format requirements are ignored in the following discussion. An Structured Query Language (SQL) like query syntax is represented as

SELECT *<attribute list>* ***FROM*** *<class names>* ***WHERE*** *< query condition>.*

Here *<attribute list>* is the list of attributes separated by commas: $Attribute_1$, $Attribute_2$, …, $Attribute_n$. At least one attribute name must be specified in *<attribute list>*. Attributes that take place in *<attribute list>* are selected from the associated classes which are specified in the ***FROM*** statement. *<class names>* contains the class names separated by commas: $Class_1$, $Class_2$, …, $Class_m$, from which the attributes are selected with the ***SELECT*** statement.

Classical databases suffer from a lack of flexibility to query. The given query condition and the contents of the database are all crisp. A query is flexible if the query condition is imprecise and uncertain and/or the databases contain imprecise and uncertain information. In the fuzzy object-oriented databases, it is shown above that an object may belong to a given class with a membership degree [0, 1].

In addition, an object satisfies the given query condition also with membership degree [0, 1] because fuzzy information occur in the query condition and/or in the object. For these reasons, the query processing in the fuzzy object-oriented database model refers to such a procedure that some objects are chosen from the classes, which first belong to the classes under the given threshold and then satisfy the given condition under the given threshold. So the queries for the fuzzy object-oriented databases are threshold-based ones. The SQL-like query syntax based on the fuzzy object-oriented database model is represented as follows.

SELECT *<attribute list>* ***FROM*** *<Class*$_1$ ***WITH*** *threshold*$_1$, …, *Class*$_m$ ***WITH*** *threshold*$_m$*>* ***WHERE*** *< query condition* ***WITH*** *threshold>*

Here, *<query condition>* is a fuzzy condition and all thresholds are crisp numbers in [0, 1]. With the SQL-like queries, one can get such objects that belong to the classes under the given thresholds and also satisfy the query condition under the given thresholds at the same time. The fuzzy selection operation defined above can be used for the fuzzy queries in the fuzzy object-oriented databases. Note that the item ***WITH*** *threshold* can be omitted when the threshold is exactly 1.

Assume we have a fuzzy class *Young Students* as follows.

CLASS *Young Students* WITH DEGREE OF *1.0*
INHERITS *Students* WITH DEGREE OF *1.0*
ATTRIBUTES
ID: TYPE OF *string* WITH DEGREE OF *1.0*
Name: TYPE OF *string* WITH DEGREE OF *1.0*
Age: FUZZY DOMAIN {*very young, young, old, very old*}: TYPE OF *integer* WITH DEGREE OF *1.0*
Height: DOMAIN *[60, 210]*: TYPE OF *real* WITH DEGREE OF *1.0*
Membership_Attribute name
WEIGHT
w (*ID*) = *0.1*
w (*Name*) = *0.1*
w (*Age*) = *0.9*
w (*Height*) = *0.2*
METHODS
…
END

And a SQL-like fuzzy query based on the class may be

SELECT *Young Students.Height* ***FROM*** *Young Students* ***WITH*** *0.5* ***WHERE*** *Young Students.Age = very young* ***WITH*** *0.8.*

This query is to get some objects of young students from the class. These objects first must belong to the class with the membership degrees equal to or greater than 0.5, and then must have very young age with a membership degree equal to or greater than 0.8. The height values of the selected objects are finally provided to the users. Assume that we now have three objects of *Young Students*: o_1, o_2 and o_3 and they have membership degrees 0.4, 0.6, and 0.7, respectively. It is clear that for the query above, o_1 does not satisfy the query because its membership degree to *Young Students* is 0.4, which is less than the given threshold 0.5. Objects o_2 and o_3 may or may not satisfy the query, depending on if their ages are very young under the given threshold 0.8.

An important issue in database queries is the complexity of queries. Among the algebraic operations, fuzzy join is an important and expensive one, and its efficient evaluation is more difficult than that of an ordinary join (Zhang and Wang 2000). So in the following, we discuss the time complexity of a SQL-like fuzzy equi-join query in a fuzzy object-oriented database. The computation of the semantic equivalence degree is the key to the meaning of the fuzzy equi-join. Suppose that we would determine if o_i [A_m] and o_j [A_m] are equivalent. Here o_i and o_j are two fuzzy objects and A_m is an attribute. We can identify two kinds of attribute A_m: simple one and complex one. For simple attribute A_m, let the universe of discourse contain n simple values. Then the time complexity of computing if o_i [A_m] and o_j [A_m] are equivalent should be $O(n^2)$ at most. For complex attribute A_m, let the universe of discourse contain n complex values and each complex value contains m simple values at most. Then the time complexity of computing if o_i [A_m] and o_j [A_m] are equivalent should be $O(n^2m^2)$ at most. It is clear that the time complexity of a SQL-like fuzzy equi-join query in a fuzzy object-oriented database mainly depends on object composition in the corresponding attributes. It should be noted that in the time complexity analysis above, we assume that the query is a flat one. It is possible that a SQL-like fuzzy equi-join query in a fuzzy object-oriented database is a nested query. At this point, the time complexity of a SQL-like fuzzy equi-join query in a fuzzy object-oriented database is also influenced by the concrete nested composition in addition to the object composition in the corresponding attributes.

For the fuzzy object-oriented databases, the contents of the databases may be fuzzy and the query conditions may be fuzzy also. So the queries only make a qualitative distinction between the returned objects and the ignored objects in the classes. The ignored objects do not satisfy the given query conditions definitely and the returned objects satisfy the given query conditions indefinitely. Here, a problem exists in fuzzy queries, i.e., the strength of query answers to the queries is unknown. Such information is very useful for ranking the query answers.

2.6 Summary

In real-world applications, information is often imprecise or uncertain. For modelling fuzzy information in the area of databases, Zadeh's fuzzy logic (Zadeh 1965) is introduced into databases to enhance the classical databases such that uncertain and imprecise information can be represented and manipulated. This resulted in numerous contributions, mainly with respect to the popular fuzzy conceptual data models and fuzzy logical database models. In this chapter, we mainly introduce several popular fuzzy database models, including fuzzy UML conceptual data model, fuzzy relational database model and fuzzy object-oriented database model. Based on the widespread studies and the relatively mature techniques of fuzzy databases, it is not surprising that fuzzy databases have been the key means for providing some technique supports for managing fuzzy data.

Besides the conceptual and logical database models, with the popularity of Web-based applications, the requirement has been put on the exchange and share of data over the Web. The XML (eXtensiable Markup Language) has become the de facto standard for data description and exchange between various systems and databases over the Internet. However, XML is not able to represent and process imprecise and uncertain data. On this basis, topics related to the modelling of fuzzy data can be considered very interesting in XML as will be introduced in the following chapter.

References

Bosc P, Prade H (1993) An introduction to fuzzy set and possibility theory based approaches to the treatment of uncertainty and imprecision in database management systems. Proceedings of the second workshop on uncertainty management in information systems: from needs to solutions

Buckles B, Petry F (1982) A fuzzy representation for relational databases. Fuzzy Sets Syst 7:213–226

Berzal F, Marín N, Pons O, Vila MA (2007) Managing fuzziness on conventional object-oriented platforms. Int J Intell Syst 22(7):781–803

Bordogna G, Pasi G (2001) Graph-based interaction in a fuzzy object oriented database. Int J Intell Syst 16:821–841

Cao TH, Rossiter JM (2003) A deductive probabilistic and fuzzy object-oriented database language. Fuzzy Sets Syst 140:129–150

Cross V, Caluwe R, Vangyseghem N (1997) A perspective from the fuzzy object data management group (FODMG). Proceedings of fuzzy systems, pp 721–728

Cross V (2001) Fuzzy extensions for relationships in a generalized object model. Int J Intell Syst 16:843–861

Chen GQ, Kerre EE (1998) Extending ER/EER concepts towards fuzzy conceptual data modeling. Proceedings of the 1998 IEEE international conference on fuzzy systems, vol 2. pp 1320–1325

Chen GQ (1999) Fuzzy logic in data modeling; semantics, constraints, and database design. Kluwer Academic Publisher, Dordrecht

Chen GQ, Vandenbulcke J, Kerre EE (1992) A general treatment of data redundancy in a fuzzy relational data model, J. Am Soc Inf Sci 43:304–311

Chaudhry N, Moyne J, Rundensteiner EA (1999) An extended database design methodology for uncertain data management. Inf Sci 121(1–2):83–112

Codd EF (1986) Missing information (applicable and inapplicable) in relational databases. SIGMOD Record 15:53–78

Codd EF (1987) More commentary on missing information in relational databases (applicable and inapplicable information). SIGMOD Record 16(1):42–50

Cubero JC, Marín N, Medina JM, Pons O, Vila MA (2004) Fuzzy object management in an object-relational framework. Proceedings of the 10th international conference on information processing and management of uncertainty in knowledge-based systems, IPMU'2004, pp 1767–1774

Dubois D, Prade H, Rossazza JP (1991) Vagueness, typicality, and uncertainty in class hierarchies. Int J Intell Syst 6:167–183

de Tré G, de Caluwe R (2003) Level-2 fuzzy sets and their usefulness in object-oriented database modeling. Fuzzy Sets Syst 140:29–49

De SK, Biswas R, Roy AR (2001) On extended fuzzy relational database model with proximity relations. Fuzzy Sets Syst 117:195–201

de Caluwe R (1998) Fuzzy and uncertain object-oriented databases: concepts and models. World Scientific Publishing Company, Singapore

Dalvi N, Suciu D (2007) Management of probabilistic data: foundations and challenges. Proceedings of the ACM SIGACT-SIGMOD-SIGART symposium on principles of database systems, pp 1–12

Gottlob G, Zicari R (1988) Closed world databases opened through null values. Proceedings of the 1988 international conference on very large data bases, pp 50–61

Grant J (1979) Partial values in a tabular database model. Inf Process Lett 9(2):97–99

Gyseghem NV, Caluwe RD (1995) Fuzzy behavior and relationships in a fuzzy OODB-model. Proceedings of the tenth annual ACM symposium on applied computing, Nashville, TN, pp 503–507

George R, Srikanth R, Petry FE, Buckles BP (1996) Uncertainty management issues in the object-oriented data model. IEEE Trans Fuzzy Syst 4(2):179–192

Haroonabadi A, Teshnehlab M (2007) Applying fuzzy-UML for uncertain systems modeling. Proceedings of the first joint congress on fuzzy and intelligent systems, Ferdowsi University of Mashhad, Iran

Haroonabadi A, Teshnehlab M (2009) Behavior modeling in uncertain information systems by Fuzzy-UML. Int J Soft Comput 4(1):32–38

Koyuncu M, Yazici A (2003) IFOOD: an intelligent fuzzy object-oriented database architecture. IEEE Trans Knowl Data Eng 15(5):1137–1154

Lee J, Xue NL, Hsu KH, Yang SJH (1999) Modeling imprecise requirements with fuzzy objects. Inf Sci 118:101–119

Liu WY, Song N (2001) The fuzzy association degree in semantic data models. Fuzzy Sets Syst 117(2):203–208

Motor A (1990) Accommodation imprecision in database systems: issues and solutions. ACM SIGMOD Record 19(4):69–74

Motor A, Smets P (1997) Uncertainty management in information systems: from needs to solutions. Kluwer Academic Publishers, Dordrecht

Ma ZM, Zhang WJ, Ma WY (2000) Semantic measure of fuzzy data in extended possibility-based fuzzy relational databases. Int J Intell Syst 15:705–716

Ma ZM, Mili F (2002) Handling fuzzy information in extended possibility-based fuzzy relational databases. Int J Intell Syst 17(10):925–942

Ma ZM, Zhang WJ, Ma WY (2004) Extending object-oriented databases for fuzzy information modeling. Inf Syst 29(5):421–435

Ma ZM (2005a) Advances in fuzzy object-oriented databases: modeling and applications. Idea Group Publishing, Hershey

Ma ZM (2005b) Fuzzy database modeling with XML (the Kluwer international series on advances in database systems). Springer, New York

Ma ZM, Yan L (2007) Fuzzy XML data modeling with the UML and relational data models. Data Knowl Eng 63(3):970–994

Ma ZM, Yan L (2008) A literature overview of fuzzy database models. J Inf Sci Eng 24(1):189–202
Ma ZM, Yan L (2010) A Literature overview of fuzzy conceptual data modeling. J Inf Sci Eng 26:427–441
Ma ZM, Zhang F, Yan L (2011) Fuzzy information modeling in UML class diagram and relational database models. Appl Soft Comput 11(6):4236–4245
Marín N, Vila MA, Pons O (2000) Fuzzy types: a new concept of type for managing vague structures. Int J Intell Syst 15:1061–1085
Marín N, Pons O, Vila MA (2001) A strategy for adding fuzzy types to an object oriented database system. Int J Intell Syst 16:863–880
Majumdar AK, Bhattacharya I, Saha AK (2002) An object-oriented fuzzy data model for similarity detection in image databases. IEEE Trans Knowl Data Eng 14:1186–1189
Ndouse TD (1997) Intelligent systems modeling with reusable fuzzy objects. Int J Intell Syst 12:137–152
Nam M, Ngoc NTB, Nguyen H, Cao TH (2007) FPDB40: a fuzzy and probabilistic object base management system. Proceedings of the FUZZ-IEEE 2007, pp 1–6
Ozgur NB, Koyuncu M, Yazici A (2009) An intelligent fuzzy object-oriented database framework for video database applications. Fuzzy Sets Syst 160:2253–2274
Petry FE (1996) Fuzzy databases: principles and applications. Kluwer Academic Publisher, Dordrecht
Parsons S (1996) Current approaches to handling imperfect information in data and knowledge bases. IEEE Trans Knowl Data Eng 8:353–372
Prade H, Testemale C (1984) Generalizing database relational algebra for the treatment of incomplete or uncertain information and vague queries. Inf Sci 34:115–143
Rundensteiner E, Bic L (1992) Evaluating aggregates in possibilistic relational databases. Data Knowl Eng 7:239–267
Raju K, Majumdar A (1988) Fuzzy functional dependencies and lossless join decomposition of fuzzy relational database systems. ACM TODS 13(2):129–166
Rundensteiner E, Hawkes LW, Bandler W (1989) On nearness measures in fuzzy relational data models. Int J Approx Reason 3:267–298
Smets P (1997) Imperfect information: imprecision-uncertainty, uncertainty management in information systems: from needs to solutions. Kluwer Academic Publishers, Dordrecht, pp 225–254
Sicilia MA, Garcia E, Gutierrez JA (2002) Integrating fuzziness in object oriented modeling language: towards a fuzzy-UML. Proceedings of international conference on fuzzy sets theory and its applications, pp 66–67
Shenoi S, Melton A (1999) Proximity relations in the fuzzy relational database model. Fuzzy Sets Syst (Suppl) 100:51–62
Umano M, Fukami S (1994) Fuzzy relational algebra for possibility-distribution-fuzzy-relational model of fuzzy data. J Intell Inf Syst 3:7–27
Umano M, Imada T, Hatono I, Tamura H (1998) Fuzzy object-oriented databases and implementation of its SQL-type data manipulation language. Proceedings of the 7th IEEE international conference on Fuzzy Systems, pp 1344–1349
Yazici A, George R (1999) Fuzzy database modeling. Physica-Verlag, Wurzburg
Yazici A, Koyuncu M (1997) Fuzzy object-oriented database modeling coupled with fuzzy logic. Fuzzy Sets Syst 89:1–26
Yan L, Ma ZM (2012) Comparison of entity with fuzzy data types in fuzzy object-oriented databases. Integr Comput-Aided Eng 19(2):199–212
Yan L, Ma ZM, Zhang F (2012) Algebraic operations in fuzzy object-oriented databases. Information systems frontiers, pp 1–14

Zvieli A, Chen PP (1986) Entity-relationship modeling and fuzzy databases. Proceedings of the 1986 IEEE international conference on data engineering, pp 320–327
Zadeh LA (1965) Fuzzy sets. Inf Control 8(3):338–353
Zadeh LA (1978) Fuzzy sets as a basis for a theory of possibility. Fuzzy Sets Syst 1(1):3–28
Zadeh LA (1975) The concept of a linguistic variable and its application to approximate reasoning. Inf Sci 8:119–249, 301–357; 9:43–80
Zaniolo C (1984) Database relations with null values. JCSS 21(1):142–162
Zicari R, Milano P (1990) Incomplete information in object-oriented databases. ACM SIGMOD Record 19(3):5–16
Zhang WN, Wang K (2000) An efficient evaluation of a fuzzy equi-join using fuzzy equality indicators. IEEE Trans Knowl Data Eng 12(2):225–237

Chapter 3
Fuzzy XML Data Models

Abstract Information is often imprecise and uncertain in many real-world applications, and thus fuzzy data modeling has been extensively investigated in various database models as introduced in Chap. 2. Currently, huge amounts of electronic data are available on the Internet, and XML has been the de facto standard of information representation and exchange over the Web. Unfortunately, XML is not able to represent and process imprecise and uncertain data. To represent and manage the imprecise and uncertain data, Zadeh's fuzzy set theory has been introduced into XML to extend XML such that uncertain and imprecise information can be represented and manipulated. The extended XML model together with the fuzzy database models introduced in Chap. 2 are the key techniques for managing fuzzy data. In this chapter, we mainly introduce fuzzy XML data model.

3.1 Introduction

As introduced in Chap. 2, fuzzy data modeling has been extensively investigated in various database models. However, the fuzzy conceptual data models and the fuzzy logical database models are not enough to represent and handle the huge amounts of electronic data which are available on the Internet. With the wide utilization of the Web and the availability of huge amounts of electronic data, information representation and exchange over the Web becomes important, and eXtensible Markup Language (XML) has been the de facto standard (Bray et al. 1998). XML and related standards are technologies that allow the easy development of applications that exchange data over the Web such as e-commerce (EC) and supply chain management (SCM). Unfortunately, although it is the current standard for data representation and exchange over the Web, XML is not able to represent and process imprecise and uncertain data. In fact, the fuzziness in EC and SCM has received considerable attentions and fuzzy set theory has been used to implement web-based business intelligence. Therefore, topics related to the

L. Yan et al., *Fuzzy XML Data Management*, Studies in Fuzziness and Soft Computing 311, DOI: 10.1007/978-3-642-44899-7_3,

modeling of fuzzy data can be considered very interesting in the XML data context. Regarding modeling fuzzy information in XML, Turowski and Weng (2002) extended XML DTDs with fuzzy information to satisfy the need of information exchange. Lee and Fanjiang (2003) studied how to model imprecise requirements with XML DTDs and developed a fuzzy object-oriented modeling technique schema based on XML. Ma and Yan (2007) and Ma (2005) proposed a fuzzy XML model for representing fuzzy information in XML documents. Tseng et al. (2005) presented an XML method to represent fuzzy systems for facilitating collaborations in fuzzy applications. Moreover, aimed at modeling fuzzy information in XML Schemas, Gaurav and Alhajj (2006) incorporated fuzziness in an XML document extending the XML Schema associated to the document and mapped fuzzy relational data into fuzzy XML. In detail, Oliboni and Pozzani (2008) proposed an XML Schema definition for representing different aspects of fuzzy information. Kianmehr et al. (2010) described a fuzzy XML schema model for representing a fuzzy relational database. In addition, XML with incomplete information (Abiteboul et al. 2001) and probabilistic data in XML (Nierman and Jagadish 2002; Senellart and Abiteboul 2007) were presented in research papers.

In this chapter, we introduce some basic notions of fuzzy XML models, including fuzziness in XML documents, fuzzy XML representation model, and fuzzy XML algebraic operations (Ma and Yan 2007; Ma et al. 2010).

3.2 Fuzziness in XML Documents

Being similar to classical XML documents, *the main part of a fuzzy XML document* consists of a *start tag*, the matching *end tag*, and everything in between is called an *element*, which can have associated *attributes*. An element may be a leaf element or a non-leaf element.

The fuzziness in an XML document is similar with the fuzziness in a relational database. In a fuzzy relational database, there may be two kinds of fuzziness as introduced in Sect. 2.4: one is *the fuzziness in tuples*, i.e., a possibility degree associated with a tuple represents the possibility of the tuple being a member of the corresponding relation; the other is *the fuzziness in attributes*, i.e., we do not know the crisp value of an attribute, and the value of the attribute may be represented by a possibility distribution.

Similar to the fuzziness in a relational database, in a fuzzy XML document, two kinds of fuzziness occur:

- *the fuzziness in elements*, and using membership degrees associated with such elements. The membership degree associated with an element represents the possibility of this element (including itself and the sub-elements rooted at it) belonging to its parent element. Now let us interpret what a membership degree associated with an element means, given that the element can nest under other elements, and more than one of these elements may have an associated membership degree. The existential membership degree associated with an element should be the possibility that the state of the world includes this element and the sub-tree rooted at it. For an element with the sub-tree rooted at it, each node in the sub-tree is not treated as independent but dependent upon its root-to-node chain. Each possibility in the source XML document is assigned conditioned on the fact that the parent element exists certainly. In other words, this possibility is a relative one based upon the assumption that the possibility the parent element exists is exactly 1.0. In order to calculate the absolute possibility, we must consider the relative possibility in the parent element. In general, the absolute possibility of an element *e* can be obtained by multiplying the relative possibilities found in the source XML along the path from *e* to the root. Of course, each of these relative possibilities will be available in the source XML document. By default, relative possibilities are therefore regarded as 1.0. Consider a chain $A \rightarrow B \rightarrow C$ from the root node *A*. Assume that the source XML document contains the relative possibilities *Poss* (*C*|*B*), *Poss* (*B*|*A*), and *Poss* (*A*), associated with the nodes *C*, *B*, and *A*, respectively. Then we have $Poss\ (B) = Poss\ (B|A) \times Poss\ (A)$ and $Poss\ (C) = Poss\ (C|B) \times Poss\ (B|A) \times Poss\ (A)$. Here, *Poss* (*C*|*B*), *Poss* (*B*|*A*), and *Poss* (*A*) can be obtained from the source XML document.

- *the fuzziness in attribute values of elements*, and using possibility distributions to represent the values of the attributes. Furthermore, attributes are classified into two types:

  single value attributes: some data items are known to have a single unique value, e.g., the age of a person in years is a unique integer, and if such a value is unknown so far, we can use the following possibility distribution: {23/0.9, 25/0.7, 27/0.6}. This is called *disjunctive* possibility distribution.
  multiple value attributes: XML restricts attributes to a single value, but it is often the case that some data item is known to have multiple values-these values may be unknown completely and can be specified with a possibility distribution. *For example*, the e-mail address of a person may be multiple character strings because he or she has several e-mail addresses available simultaneously. In case we do not have complete knowledge of the e-mail address for "Tom Smith", we may say that the e-mail address may be "TSmith@yahoo.com" with possibility 0.6 and "Tom_Smith@hotmail.com" with possibility 0.85. This is called *conjunctive* possibility distribution.

For ease of understanding, we interpret the above two kinds of fuzziness with a simple *fuzzy XML document* d_1 in Fig. 3.1. In Fig. 3.1, we talk about the *universities* in an area of a given city, say, *Detroit*, *Michigan*, in the USA.

(a) *Wayne State University* is located in downtown *Detroit*, and thus the possibility that it is included in the universities in *Detroit* is 1.0. For pair <Val Poss = 1.0> ··· </Val> is omitted (see *Lines* 50–51).
(b) *Oakland University*, however, is located in a *nearby county* of *Michigan*, named *Oakland*. Whether *Oakland University* is included in the universities in *Detroit* depends on how to define the area of *Detroit*, *the Greater Detroit Area*

```
1.  <universities>
2.    <university UName = "Oakland University">
3.      <Val Poss = 0.8>
4.        <department DName = "Computer Science and Engineering">
5.          <employee FID = "85431095">
6.            <Dist type = "disjunctive">
7.              <Val Poss = 0.8>
8.                <fname>Frank Yager</name>
9.                <position>Associate Professor</position>
10.               <office>B1024</office>
11.               <course>Advances in Database Systems</course>
12.             </Val >
13.             <Val Poss = 0.6>
14.               <fname>Frank Yager</name>
15.               <position>Professor</position>
16.               <office>B1024</office>
17.               <course>Advances in Database Systems</course>
18.             </Val >
19.           </Dist>
20.         </employee>
21.         <student SID = "96421027">
22.           <sname>Tom Smith</name>
23.           <age>
24.             <Dist type = "disjunctive">
25.               <Val Poss = 0.4>23</Val>
26.               <Val Poss = 0.6>25</Val>
27.               <Val Poss = 0.8>27</Val>
28.               <Val Poss = 1.0>29</Val>
29.               <Val Poss = 1.0>30</Val>
30.               <Val Poss = 1.0>31</Val>
31.               <Val Poss = 0.8>33</Val>
32.               <Val Poss = 0.6>35</Val>
33.               <Val Poss = 0.4>37</Val>
34.             </Dist>
35.           </age>
36.           <sex>Male</sex>
37.           <email>
38.             <Dist type= "conjunctive">
39.               <Val Poss = 0.60>TSmith@yahoo.com</Val>
40.               <Val Poss = 0.85>Tom_Smith@yahoo.com</Val>
41.               <Val Poss = 0.85>Tom_Smith@hotmail.com</Val>
42.               <Val Poss = 0.55>TSmith@hotmail.com</Val>
43.               <Val Poss = 0.45>TSmith@msn.com</Val>
44.             </Dist>
45.           </email>
46.         </student>
47.       </department >
48.     </Val>
49.   </university>
50.   <university Uname = "Wayne State University">
51.   </university>
52. </universities >
```

Fig. 3.1 A fragment of a fuzzy XML document d_1

or *only the City of Detroit*. Assume that it is unknown and the possibility that *Oakland University* is included in the universities in *Detroit* is assigned 0.8 (see *Line* 3). The cases 1–2 are *the fuzziness in elements*. The degree associated with such an element represents the possibility that a university is included in universities in *Detroit*.

(c) For the student *Tom Smith*, if his age is unknown so far, i.e., he has fuzzy value in the attribute *age*. Since *age* is known to have a single unique value, we can use the *disjunctive* possibility distribution to represent such value (see *Lines* 23–35).

(d) The *e-mail* address of *Tom Smith* may be multiple character strings because he has several e-mail addresses simultaneously. If we do not know his exact e-mail addresses, and we use the *conjunctive* possibility distribution to represent such information and may say that the e-mail address may be "TSmith@yahoo.com" with possibility 0.6 and "TSmith@msn.com" with possibility 0.45 (see *Lines* 37–45). *Note that*, the cases 3–4 are *the fuzziness in attribute values of elements*. In an XML document, it is often the case that some values of attributes may be unknown completely and can be specified with possibility distributions.

3.3 Fuzzy XML Representation Models and Formalizations

In the following, we introduce fuzzy XML representation models, including the representation of fuzzy data in the XML document, and two fuzzy XML document structures fuzzy DTD and fuzzy XML Schema.

3.3.1 Representation of Fuzzy Data in the XML Document

In order to represent the fuzzy data in XML documents, it is shown in the previous part that several fuzzy constructs (such as *Poss*, *Val* and *Dist*) are introduced:

1. A possibility attribute called ***Poss***, which takes a value of [0, 1], is introduced. The attribute *Poss* is applied together with a fuzzy construct ***Val*** to specify the possibility of an element existing in the fuzzy XML document (see *Line* 3 in Fig. 3.1).
2. Another fuzzy construct called ***Dist***, which specifies a possibility distribution, is introduced. Based on pair <Val Poss> and </Val>, possibility distribution for an element can be expressed. Also, possibility distribution can be used to express fuzzy element values. For this purpose, we introduce another fuzzy construct called Dist to specify a possibility distribution. Typically, a Dist element has multiple Val elements as children, each with an associated possibility. Since we have two types of possibility distribution, the Dist construct should indicate the type of a possibility distribution being disjunctive or conjunctive. (see *Lines* 24–34 and *Lines* 38–44 in Fig. 3.1)

Again consider Fig. 3.1. Lines 24–34 are the disjunctive Dist construct for the age of student "Tom Smith". Lines 38–44 are the conjunctive Dist construct for the email of student "Tom Smith". It should be noted, however, that the possibility distributions in Lines 24–34 and *Lines* 38–44 are all for leaf nodes in the ancestor–descendant chain. In fact, we can also have possibility distributions and values over non-leaf nodes. Observe the disjunctive Dist construct in Lines 6–19, which express the two possible statuses for the employee with ID 85431095. In these two employee values, Lines 7–12 are with possibility 0.8, and Lines 13–18 with possibility 0.6.

The structure of an XML document can be described by *Document Type Definition* (*DTD*) or *XML Schema* (Antoniou and van Harmelen 2004). A DTD, which defines the valid elements and their attributes and the nesting structures of these elements in the instance documents, is used to assert the set of "rules" that each instance document of a given document type must conform to. XML Schemas provide a much more powerful means than DTDs by which to define your XML document structure and limitations. It has been shown that the XML document must be extended for fuzzy data modeling. As a result, several fuzzy constructs have been introduced. In order to accommodate these fuzzy constructs, it is clear that the DTD or XML Schema of the source XML document should be correspondingly modified. In the following sub-chapters, we focus on DTD and XML Schema modification for fuzzy XML data modeling.

3.3.2 Fuzzy DTD

In the following, we define DTD modification (i.e., fuzzy DTD) for representing the structure of the fuzziness in XML document as introduced in Sect. 3.3.1.

Firstly, we define the ***basic elements*** in a fuzzy DTD as follows:

<!ELEMENT *element*$_1$ (*element*$_2$?, *, +)>

//*element*$_1$ contains *element*$_2$, and the appearance times of *element*$_2$ are restricted by the cardinalities: ? denotes 0 or 1 time; * denotes 0 or *n* times; + denotes 1 or *n* times; No cardinality operator means exactly once

<!ELEMENT *element*$_1$ (*element*$_2$, *element*$_3$, …)>

//*element*$_1$ contains *element*$_2$, *element*$_3$, … in order

<!ELEMENT *element*$_1$ (*element*$_2$ | *element*$_3$ | …)>

//*element*$_1$ contains either *element*$_2$ or *element*$_3$, …

<!ELEMENT *element*$_1$ (#PCDATA)>

//#PCDATA, which is the only atomic type for elements, denotes *element*$_1$ may have any content

<!ELEMENT *element*$_1$ (empty)>

//*element*$_1$ is an empty element

Moreover, the *attributes* of an element $element_i$ can be represented as follows:

<!ELEMENT $element_i$ …>

<!ATTLIST $element_i$ AttName AttType ValType>

Here AttName is the name of the attribute, AttType is the type of the attribute, and ValType is the value type which can be #REQUIRED, #IMPLIED, #FIXED "value", and "value" (Antoniou and van Harmelen 2004).

Then, we define ***Val*** and ***Dist*** elements as follows:

<!ELEMENT Val (#PCDATA | *basic_definition*)>

<!ATTLIST Val Poss CDATA "1.0">

//basic_definition represents any case of the basic element definitions above

<!ELEMENT Dist (Val+)>

<!ATTLIST Dist type (disjunctive|conjunctive) "disjunctive">

Finally, based on the *Val* and *Dist* elements, we modify the basic element definitions above so that all of the elements can use possibility distributions (Dist). In summary, the basis elements can be classified into two types, i.e., the *leaf element* and the *non-leaf element*:

- for the *leaf element* which only contains #PCDATA, say leafElement, its definition is modified from <!ELEMENT leafElement (#PCDATA)> to

  <!ELEMENT leafElement (#PCDATA | Dist)>.

  That is, a leaf element may be fuzzy and takes a value represented by a possibility distribution.
- for the *non-leaf element* which contains the other elements, say nonleafElement, its definition is modified from <!ELEMENT nonleafElement (*basic_definition*)> to

  <!ELEMENT nonleafElement (*basic_definition* | Val+ | Dist)>

  That is, a non-leaf element may be crisp, e.g., *student* in Fig. 3.1, and thus the non-leaf element *student* can be defined as

  <!ELEMENT student (sname?, age?, sex?, email?)>.

  Also, a non-leaf element may be fuzzy and takes a value represented by a possibility distribution. We differentiate two cases: *the first one is the element takes a value connected with a possibility degree*, e.g., *university* in Fig. 3.1, which can be defined as

  <!ELEMENT university (Val+)>

  and *the second one is the element takes a set of values and each value is connected with a possibility degree*, e.g., *age* of student in Fig. 3.1, which can be defined as

  <!ELEMENT age (Dist)>.

Based on the above modified fuzzy DTD definitions, Fig. 3.2 gives the *fuzzy DTD* D_1 w.r.t. the fuzzy XML document d_1 in Fig. 3.1.

3.3.3 Fuzzy XML Schema

Being similar to the fuzzy DTD , in the following, we define the XML Schema modification (i.e., fuzzy XML Schema) for representing the structure of the fuzziness in XML document as introduced in Sect. 3.3.1.

```
<!ELEMENT universities (university*)>
<!ELEMENT university (Val+)>
    <!ATTLIST university UName IDREF #REQUIRED>
<!ELEMENT Val (department*)>
    <!ATTLIST Val Poss CDATA "1.0">
<!ELEMENT department (employee*, student*)>
    <!ATTLIST department DName IDREF #REQUIRED>
<!ELEMENT employee (Dist)>
    <!ATTLIST employee FID IDREF #REQUIRED>
<!ELEMENT Val (fname?, position?, office?, course?)>
    <!ATTLIST Val Poss CDATA "1.0">
<!ELEMENT student (sname?, age?, sex?, email?)>
    <!ATTLIST student SID IDREF #REQUIRED>
<!ELEMENT fname (#PCDATA)>
<!ELEMENT position (#PCDATA)>
<!ELEMENT office (#PCDATA)>
<!ELEMENT course (#PCDATA)>
<!ELEMENT sname (#PCDATA)>
<!ELEMENT age (Dist)>
<!ELEMENT Dist (Val+)>
    <!ATTLIST Dist type (disjunctive)>
<!ELEMENT sex (#PCDATA)>
<!ELEMENT email (Dist)>
<!ELEMENT Dist (Val+)>
    <!ATTLIST Dist type (conjunctive)>
<!ELEMENT Val (#PCDATA)>
    <!ATTLIST Val Poss CDATA "1.0">
```

Fig. 3.2 The fuzzy DTD D_1 w.r.t. the fuzzy XML document d_1 in Fig. 3.1

First we define ***Val*** element as follows:

```
<xs:element name="Val" type="valtype"/>
    <xs:complexType name="valtype">
    <xs:sequence>
        <xs:element name="original-definition" minOccurs="0" maxOccurs=
        "unbounded"/>
        <xs:attribute name="Poss" type="xs:fuzzy" minOccurs="0" maxOccurs=
        "unbounded" default="1.0"/>
    </xs:sequence>
    </xs:complexType>
```

Then we define ***Dist*** element as follows:

```
<xs:element name="Dist" type="disttype"/>
<xs:complexType name="disttype">
<xs:element name="Val" type="valtype" minOccurs="1" maxOccurs="unbounded"/>
<xs:attribute values="disjunctive conjunctive" default="disjunctive"/>
</xs:complexType>
```

Now we modify the element definition in the classical Schema so that all of the elements can use possibility distributions (Dist). For a sub-element that only contains leaf elements, its definition in the Schema is as follows.

```
<xs:element name="leafElement" type="leafelementtype"/>
<xs:complexType name="leafelementtype">
<xs:sequence>
    <xs:element name="original-definition" type="xs:type" minOccurs="0"
    maxOccurs="unbounded"/>
    <xs:element name="Dist" type="disttype" minOccurs="0" maxOccurs=
    "unbounded"/>
</xs:sequence>
</xs:complexType>
```

For an element that contains leaf elements without any fuzziness, its definition in the Schema is as follows.

```
<xs:element name="original-definition" type="xs:type" minOccurs="0"
maxOccurs="unbounded"/>
```

For an element that contains leaf elements with fuzziness, its definition in the Schema is as follows.

```
<xs:element name="leafElement" type="leafelementtype"/>
<xs:complexType name="leafelementtype">
```

```
<xs:element name="Dist" type="disttype" minOccurs="0" maxOccurs=
"unbounded"/>
</xs:complexType>
```

For a sub-element that does not contain any leaf elements, its definition in the Schema is as follows.

```
<xs:element name="nonleafElement" type="nonleafelementtype"/>
<xs:complexType name="nonleafelementtype">
<xs:sequence>
    <xs:element name="original-definition" type="xs:type" minOccurs="0"
    maxOccurs="unbounded"/>
    <xs:element name="Dist" type="disttype" minOccurs="0" maxOccurs=
    "unbounded"/>
    <xs:element name="Val" type="valtype" minOccurs="0" maxOccurs=
    "unbounded"/>
</xs:sequence>
</xs:complexType>
```

For an element that does not contain leaf elements without any fuzziness, its definition in the Schema is as follows.

```
<xs:element name="nonleafElement" type="nonleafelementtype"/>
<xs:complexType name="nonleafelementtype">
<xs:element name="original-definition" type="xs:type" minOccurs="0"
maxOccurs="unbounded"/>
</xs:complexType>
```

For a sub-element that does not contain leaf elements but a fuzzy value, its definition in the Schema is as follows.

```
<xs:element name="nonleafElement" type="nonleafelementtype"/>
<xs:complexType name="nonleafelementtype">
<xs:element name="Val" type="valtype" minOccurs="0" maxOccurs=
"unbounded"/>
</xs:complexType>
```

For a sub-element that does not contain leaf elements but a set of fuzzy values, its definition in the Schema is as follows.

```
<xs:element name="nonleafElement" type="nonleafelementtype"/>
<xs:complexType name="nonleafelementtype">
<xs:element name="Dist" type="disttype" minOccurs="0" maxOccurs=
"unbounded"/>
</xs:complexType>
```

The fuzzy XML Schema w.r.t. the fuzzy XML document in Fig. 3.1 is shown as follows:

```
<? xml version="1.0"? >
<xs:schema xmlns:xs="http://www.w3.org/2001/XMLSchema">
<xs:element name="universities">
<xs:complexType>
<xs:element name="university" type="universityype" minOccurs="0"
maxOccurs="unbounded"/>
</xs:complexType>
</xs:element>
<xs:complexType name="universityype">
<xs:element name="Val" type="valtype" minOccurs="1"
maxOccurs="unbounded"/>
<xs:attribute name="UName" type="xs:IDREF"
use="REQUIRED"/>
</xs:complexType>
<xs:complexType name="valtype">
<xs:sequence>
<xs:element name="department" type="worktype" minOccurs="0"
maxOccurs="unbounded"/>
<xs:element name="fname" type="xs:string" minOccurs="0"
maxOccurs="1"/>
<xs:element name="position" type="xs:string" minOccurs="0"
maxOccurs="1"/>
<xs:element name="office" type="xs:string" minOccurs="0"
maxOccurs="1"/>
<xs:element name="course" type="xs:string" minOccurs="0"
maxOccurs="1"/>
</xs:sequence>
<xs:attribute name="Poss" type="xs:fuzzy" minOccurs="0"
maxOccurs="unbounded" default="1.0"/>
</xs:complexType>
<xs:complexType name="worktype">
<xs:sequence>
<xs:element name="employee" type="employeetype" minOccurs="0"
maxOccurs="unbounded"/>
<xs:element name="student" type="studenttype" minOccurs="0"
maxOccurs="unbounded"/>
</xs:sequence>
<xs:attribute name="DName" type="xs:IDREF"
use="REQUIRED"/>
</xs:complexType>
<xs:complexType name="employeetype">
<xs:element name="Dist" type="disttype"/>
```

```
<xs:attribute name="FID" type="xs:IDREF" use="REQUIRED"/>
</xs:complexType>
<xs:complexType name="disttype">
<xs:element name="Val" type="valtype" minOccurs="1"
maxOccurs="unbounded"/>
<xs:attribute values="disjunctive conjunctive"
default="disjunctive"/>
</xs:complexType>
<xs:complexType name="studenttype">
<xs:sequence>
<xs:element name="sname" type="xs:string" minOccurs="0"
maxOccurs="1"/>
<xs:element name="age" type="agetype" minOccurs="0"
maxOccurs="1"/>
<xs:element name="sex" type="xs:string" minOccurs="0"
maxOccurs="1"/>
<xs:element name="email" type="emailtype" minOccurs="0"
maxOccurs="1"/>
</xs:sequence>
<xs:attribute name="SID" type="xs:IDREF" use="REQUIRED"/>
</xs:complexType>
<xs:complexType name="agetype">
<xs:element name="Dist" type="disttype"/>
</xs:complexType>
<xs:complexType name="emailtype">
<xs:element name="Dist" type="disttype"/>
<xs:attribute values="conjunctive"/>
</xs:complexType>
</xs:schema>
```

3.3.4 Formalization of Fuzzy XML Models

Being similar to the classical XML document, a *fuzzy XML document* can be intuitionally seen as a *syntax tree*. Figure 3.3 shows a fragment of the fuzzy XML document d_1 in Fig. 3.1 and its tree representation.

Based on the tree representation of the fuzzy XML document, in the following we define the formalization of fuzzy XML models in Ma et al. (2010), Zhang et al. (2013).

It can be found from Fig. 3.2 that a *fuzzy DTD* is made up of *element type definitions*, and each element may have associated attributes. Each element type definition has the form $E \rightarrow (\alpha, A)$, where E is the defined element type (e.g., *university* and *student*), α, called the *content model* such as *university* (*UName*, *Val*+), and A are attributes of E.

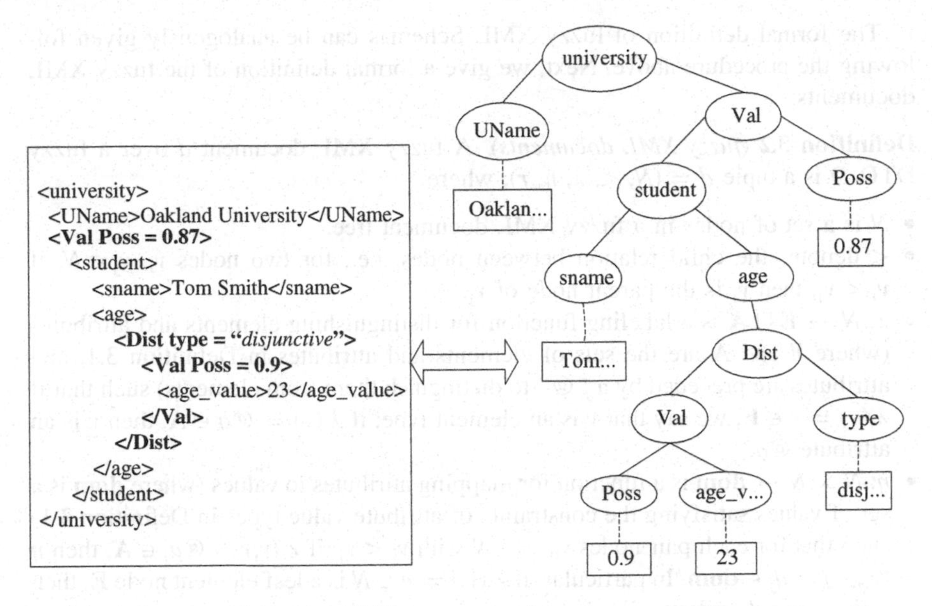
<university>
<UName>Oakland University</UName>
<Val Poss = 0.87>
<student>
<sname>Tom Smith</sname>
<age>
<Dist type = "disjunctive">
<Val Poss = 0.9>
<age_value>23</age_value>
</Val>
</Dist>
</age>
</student>
</university>
university
UName
Val
Oaklan...
student
Poss
0.87
sname
age
Tom...
Dist
Val
type
Poss
age_v...
disj...
0.9
23

Fig. 3.3 A fragment of the fuzzy XML document and its tree representation

For the sake of simplicity, we assume that the symbol **T** denotes the atomic types of elements and attributes such as #PCDATA and CDATA, **E** denotes the set of elements including the basic elements (e.g., *university* and *student*) and the special elements (e.g., *Val* and *Dist*), **A** denotes the set of attributes, and $\mathbf{S} = \mathbf{T} \cup \mathbf{E}$.

Definition 3.1 (*fuzzy DTDs*) A fuzzy DTD *D* is a pair (**P**, *r*), where **P** is a set of *element type definitions*, and $r \in \mathbf{E}$ is the *root element type*, which uniquely identifies a fuzzy DTD. Each element type definition has the form $E \rightarrow (\alpha, A)$, constructed according to the following syntax:

$$\alpha ::= S \mid \text{empty} \mid (\alpha_1|\alpha_2) \mid (\alpha_1, \alpha_2) \mid \alpha? \mid \alpha^* \mid \alpha+ \mid \text{any}$$
$$A ::= \text{empty} \mid (\text{AN}, \text{AT}, \text{VT})$$

Here:

1. $\mathbf{S} = \mathbf{T} \cup \mathbf{E}$; *empty* denotes the empty string; "|" denotes *union*, and "," denotes *concatenation*; α can be extended with cardinality operators "?", "*", and "+", where "?" denotes 0 or 1 time, "*" denotes 0 or *n* times, and "+" denotes 1 or *n* times; the construct *any* stands for any sequence of element types defined in the fuzzy DTD;
2. $AN \in \mathbf{A}$ denotes the attribute names of the element *E*; *AT* denotes the attribute types; and *VT* is the value types of attributes which can be #REQUIRED, #IMPLIED, #FIXED "value", "value", and disjunctive/conjunctive possibility distribution.

The formal definition of fuzzy XML Schemas can be analogously given following the procedure above. Next, we give a formal definition of the fuzzy XML documents.

Definition 3.2 (*fuzzy XML documents*) A fuzzy XML document d over a fuzzy DTD D is a tuple $d = (N, <, \lambda, \eta, r)$, where:

- N is a set of nodes in a fuzzy XML document tree.
- $<$ denotes the child relation between nodes, i.e., for two nodes $v_i, v_j \in N$, if $v_i < v_j$, then v_i is the parent node of v_j.
- $\lambda: N \rightarrow \mathbf{E} \cup \mathbf{A}$ is a labeling function for distinguishing elements and attributes (where **E** and **A** are the sets of elements and attributes in Definition 3.1, and attributes are preceded by a "@" to distinguish them from elements) such that if $\lambda (v) = e \in \mathbf{E}$, we say that v is an element type; if $\lambda (v) = @a \in \mathbf{A}$, then v is an attribute $@a$.
- $\eta: N \times N \rightarrow \mathbf{dom}$ is a function for mapping attributes to values (where **dom** is a set of values satisfying the constraints of attribute value types in Definition 3.1) such that for each pair nodes $v_i, v_j \in N$ with $v_i < v_j$, if $\lambda (v_j) = @a_j \in \mathbf{A}$, then $\eta (v_i, v_j) = d_j \in \mathbf{dom}$. In particular, if $\lambda (v_j) = e \in N$ is a leaf element node **E**, then $\eta (v_i, v_j) = d_j \in \mathbf{dom}$.
- r is the root node of a fuzzy XML document tree.

In the following, we further give the formalization of fuzzy XML data models, which is defined based on the characteristic of the tree structure of fuzzy XML data models as mentioned above.

In short, the basic structure of a fuzzy XML data model is a tree. Let N be a finite set (of vertices), $E \in N \times N$ be a set (of edges) and $\lambda: E \rightarrow L$ be a mapping from edges to a set L of strings called labels. The triple $G = (N, E, \lambda)$ is an edge labeled directed graph. It should be noted that the tree structure only briefly describes the characteristic of fuzzy XML data models, and ignores a number of fuzzy XML features. Here, we further provide a more detailed formal definition of fuzzy XML tree.

Definition 3.3 A fuzzy XML tree t can be a tuple $t = (N, \sigma, \lambda, \eta, \rho, \gamma, \propto)$:

- $N = \{N_1, \ldots, N_n\}$ is a set of vertices.
- $\sigma \subset \{(N_i, N_j) \mid N_i, N_j \in N\}$, (N, σ) is a directed tree.
- $\ell: N \rightarrow (L \cup \{\text{NULL}\})$, where L a set of strings called labels. For $n \in N$ and $l \in L$, $\lambda (n, l)$ specifies the set of objects that may be children of n with label l.
- $\eta \rightarrow T$, where T is a set of fuzzy XML types (Oliboni and Pozzani 2008).
- ρ is a mapping from the set of objects $v \in V$ to local possibility functions. It defines the possibility of a set of children of an object existing given that the parent object exists.
- γ associates with $n \in N$, each label $l \in L$, and an integer-valued interval function, i.e., $\gamma (n, l) = [min, max]$. γ is used to define the cardinality constrains of children with a label.

- $\propto$ is a possibly empty partial order on N. Here, a relation "$\propto$" is a partial order on a set N if the following three characteristics hold: (1) reflexivity: $\theta \propto \theta$ for all $\theta \in N$; (2) antisymmetry: $\theta \propto \omega$ and $\omega \propto \theta$ implies $\omega = \theta$; (3) transitivity: $\theta \propto \omega$ and $\omega \propto \varepsilon$ implies $\theta \propto \varepsilon$.

3.4 Fuzzy XML Algebraic Operations

As evidenced by the database management systems, a formal algebra is essential for applying database-style optimization to query processing. Similarly, along the fuzzy XML model as introduced in the previous chapters, the fuzzy XML algebraic operations should be defined for supporting fuzzy XML queries. As mentioned in Hung et al. (2003), an extension of the relational algebra, which is based on probabilistic instance names and path expressions to handle probabilistic semi-structured data, is investigated. It should be noted that the XML data are order sensitive, and their studies may be more reasonable if they had considered the order problem. In this section, we introduce a general algebraic framework for supporting imprecise and uncertain XML queries (Ma et al. 2010). The algebra serves as a target language for translation from declarative user oriented query language for fuzzy XML. It is user-friendly and can provide a concise representation of query execution. The algebra also supports order-sensitive fuzzy queries, which has been ignored in most of algebras. In particular, it is not only designed to integrate with tuple operators but also supports fuzzy tree patterns queries. In the following, we introduce several common fuzzy XML algebraic operators.

Union Given two fuzzy XML trees $t_1 = (N_1, \sigma_1, \lambda_1, \eta_1, \rho_1, \gamma_1, \propto_1)$ and $t_2 = (N_2, \sigma_2, \lambda_2, \eta_2, \rho_2, \gamma_2, \propto_2)$, and t_1 and t_2 are isomorphic, the union $\cup$ can be defined as:

- $\alpha\ (t_1 \cup t_2) = (N_1 \cup N_2, \sigma_1 \cup \sigma_2, \lambda_1 \cup \lambda_2, \eta_1 \cup \eta_2, \rho_1 \cup \rho_2, \gamma_1 \cup \gamma_2, \propto_1 \cup \propto_2)$
- $\beta\ (t_1 \cup t_2) \in \{s \mid s \in \beta\ (t_1) \text{ or } s \in \beta\ (t_2)\}$

Based on fuzzy set theory, we have $\rho_{ur} = \text{Max}\ (\rho_1, \rho_2)$, where ρ_{ur} is the membership degree of fuzzy union result.

Intersection Given two fuzzy XML trees $t_1 = (N_1, \sigma_1, \lambda_1, \eta_1, \rho_1, \gamma_1, \propto_1)$ and $t_2 = (N_2, \sigma_2, \lambda_2, \eta_2, \rho_2, \gamma_2, \propto_2)$, and t_1 and t_2 are isomorphic, the union $\cap$ can be defined as:

- $\alpha\ (t_1 \cap t_2) = (N_1 \cap N_2, \sigma_1 \cap \sigma_2, \lambda_1 \cap \lambda_2, \eta_1 \cap \eta_2, \rho_1 \cap \rho_2, \gamma_1 \cap \gamma_2, \propto_1 \cap \propto_2)$
- $\beta\ (t_1 \cap t_2) \in \{s \mid s \in \beta\ (t_1) \text{ and } s \in \beta\ (t_2)\}$

Further, we have $\rho_{ir} = \text{Min}(\rho_1, \rho_2)$, where ρ_{ir} is the membership degree of fuzzy union result.

Difference Given two fuzzy XML trees $t_1 = (N_1, \sigma_1, \lambda_1, \eta_1, \rho_1, \gamma_1, \propto_1)$ and $t_2 = (N_2, \sigma_2, \lambda_2, \eta_2, \rho_2, \gamma_2, \propto_2)$, and t_1 and t_2 are isomorphic, the union—can be defined as:

- $\alpha(t_1 - t_2) = (N_1 - N_2, \sigma_1 - \sigma_2, \lambda_1 - \lambda_2, \eta_1 - \eta_2, \rho_1 - \rho_2, \gamma_1 - \gamma_2, \propto_1 - \propto_2)$
- $\beta(t_1 - t_2) \in \{s \mid s \in \beta(t_1) \text{ and } s \notin \beta(t_2)\}$

Further, we have $\rho_{dr} = \text{Min}(\rho_1, \rho_2')$, $\rho_2' = 1 - \rho_2$ is the complement of ρ_2, where ρ_{dr} is the membership degree of fuzzy difference result.

Cartesian product Given two fuzzy XML trees $t_1 = (N_1, \sigma_1, \lambda_1, \eta_1, \rho_1, \gamma_1, \propto_1)$ and $t_2 = (N_2, \sigma_2, \lambda_2, \eta_2, \rho_2, \gamma_2, \propto_2)$, and t_1 and t_2 are isomorphic, the Cartesian product $\otimes$ can be defined as:

- $\alpha(t_1 \otimes t_2) = (N_1 \otimes N_2, \sigma_1 \otimes \sigma_2, \lambda_1 \otimes \lambda_2, \eta_1 \otimes \eta_2, \rho_1 \otimes \rho_2, \gamma_1 \otimes \gamma_2, \propto_1 \otimes \propto_2)$
- $\beta(t_1 \otimes t_2) \in \{(s_1, s_2) \mid s_1 \in \beta(t_1) \text{ and } s_2 \in \beta(t_2)\}$

Further, we have $\rho_{cr} = \text{Min}(\rho_1, \rho_2)$, where ρ_{dr} is the of fuzzy cartesian product result.

Selection Given a fuzzy XML trees $t = (N, \sigma, \lambda, \eta, \rho, \gamma, \propto)$, if there is a predicate δ, then we have the definition of Selection: $\varepsilon_\delta(t) = \{s \mid s \Leftarrow t \cap tr(\delta, s)\}$, where function $tr(\delta, s)$ is used to extract the pattern from δ.

The selection operator ε filters the fuzzy trees using a special predicate that can be any combination of logical operators and simple qualifications (specified via a pattern). It accepts a set of data trees t as input. The output is the entire set of the matching witness trees for all input trees, which is not only the content of right result, but also the structure of objective trees.

Projection Given a fuzzy XML trees $t = (N, \sigma, \lambda, \eta, \rho, \gamma, \propto)$, ζ is a fuzzy projection function, then we have the definition of Projection: $\psi_\zeta(t) = \{\zeta(s) \mid s \in t\}$.

The projection operator ψ is used to eliminate objects. In the substructure resulting from node elimination, we would expect the hierarchical relationships between surviving objects that existed in the input trees to be preserved.

Join Given two fuzzy XML trees $t_1 = (N_1, \sigma_1, \lambda_1, \eta_1, \rho_1, \gamma_1, \propto_1)$ and $t_2 = (N_2, \sigma_2, \lambda_2, \eta_2, \rho_2, \gamma_2, \propto_2)$, and a condition predicate δ, the join $\oplus$ can be defined as:

- $\alpha(t_1 \oplus_\delta t_2) = \alpha(t_1 \otimes_\delta t_2)$
- $\beta(t_1 \oplus_\delta t_2) \subseteq \{\beta_\delta(t_1 \otimes_\delta t_2)\}$

The join operator $\oplus$ is a mixed of Cartesian product and F-Selection, and joins trees on a predicate.

Grouping Given a fuzzy XML trees $t = (N, \sigma, \lambda, \eta, \rho, \gamma, \propto)$, and $g(t, \delta)$ is a function that represents sub-tree equivalence object over t under predicate δ condition, then we have the definition of the grouping ξ:

$$\xi_\delta(t) = \{ \bigcup_{s \in g(t,\delta)} s \mid s \in t\}$$

The grouping operator ξ is to split a collection into subsets of trees and represent each subset as an ordered tree in some meaningful way.

Ordering Given a fuzzy XML trees $t = (N, \sigma, \lambda, \eta, \rho, \gamma, \propto)$, if there is an ordering rule μ: $\varpi \rightarrow \omega$, then we have the definition of the ordering v:

$$\upsilon(t, \mu) = \{s \mid s \in t \; and \; \mu(s, \propto, \omega) \Leftarrow \mu(s, \propto, \varpi)\}$$

Bind Given a fuzzy XML trees $t = (N, \sigma, \lambda, \eta, \rho, \gamma, \propto)$, and a function φ (t, δ) which can extract γ from t with the restriction δ, if there is a rule : $\varpi \rightarrow \omega$, then we have the definition of the Bind π:

$$\pi(t) = \sum_{\varphi(t,\delta)} \{s \mid s \in t \; and \; \mu(s, \omega) \Leftarrow \mu(s, \varpi)\}$$

The Bind operation π is to extract information from an input tree according to a given filter, and produce a structure, which comparable to a ¬1NF relation. Moreover, the Tree operation is an inverse operation to Bind, which can be used to generate a new nested XML structure.

3.5 Summary

In real-world applications, information is often imprecise or uncertain. The fuzzy conceptual data models and fuzzy logical database models mentioned in Chap. 2 are not enough to represent and handle the huge amounts of electronic data which are available on the Internet. Currently, XML has been the de facto standard of information representation and exchange over the Web. Unfortunately, XML is not able to represent and process imprecise and uncertain data, although the databases with imprecise and uncertain information have been extensively discussed. Therefore, topics related to the modelling of fuzzy data are considered very interesting in the XML data context. In this chapter, we introduced fuzzy XML models, including fuzziness in XML documents, fuzzy XML representation model, and fuzzy XML algebraic operations

In order to meet the needs of practical application, just providing the modelling technology of fuzzy XML is not enough, fuzzy XML query is also very necessary. How to query XML with imprecise or uncertain information has raised certain concerns as will be introduced in the following chapter.

References

Abiteboul S, Segoufin L, Vianu V (2001) Representing and querying XML with incomplete information. In: Proceedings of the ACM SIGACT-SIGMOD-SIGART symposium on principles of database systems (PODS), pp 150–161

Antoniou G, van Harmelen F (2004) A semantic web primer. MIT Press, Cambridge

Bray T, Paoli J, Sperberg-McQueen CM (eds) (1998) Extensible markup language (XML) 1.0, W3C Recommendation. http://www.w3.org/TR/1998/REC-xml-19980210

Gaurav A, Alhajj R (2006) Incorporating fuzziness in XML and mapping fuzzy relational data into fuzzy XML. In: Proceedings of ACM symposium on applied computing, ACM, pp 456–460

Hung E, Getoor L, Subrahmanian VS (2003) PXML: a probabilistic semistructured data model and algebra. In: Proceedings of the 19th international conference on data engineering (ICDE'03), pp 467–478

Kianmehr K, Özyer T, Lo A, Jiwani A, Alimohamed Y, Spence K, Alhajj R (2010) Human centric data representation: from fuzzy relational databases into fuzzy XML. In: Ma Z (ed) Soft computing in XML data management. Springer, Berlin

Lee J, Fanjiang Y (2003) Modeling imprecise requirements with XML. Inf Softw Technol 45(7):445–460

Ma ZM (2005) Fuzzy database modeling with XML (the Kluwer international series on advances in database systems). Springer, New York

Ma ZM, Yan L (2007) Fuzzy XML data modeling with the UML and relational data models. Data Knowl Eng 63(3):970–994

Ma ZM, Liu J, Yan L (2010) Fuzzy data modeling and algebraic operations in XML. Int J Intell Syst 25(9):925–947

Nierman A, Jagadish HV (2002) ProTDB: probabilistic data in XML. In: Proceedings of the VLDB 2002, pp 646–657

Oliboni B, Pozzani G (2008) Representing fuzzy information by using XML schema. In: Proceedings of database and expert systems applications (DEXA 2008), pp 683–687

Senellart P, Abiteboul S (2007) On the complexity of managing probabilistic XML data. In: Proceedings of the ACM SIGACT-SIGMOD-SIGART symposium on principles of database systems (PODS), pp 283–292

Tseng C, Khamisy W, Vu T (2005) Universal fuzzy system representation with XML. Comput Stand Interfaces 28(2):218–230

Turowski K, Weng U (2002) Representing and processing fuzzy information-an XML-based approach. J Knowl Based Syst 15(1–2):67–75

Zhang F, Ma ZM, Yan L (2013) Construction of fuzzy ontologies from fuzzy XML models. Knowl Based Syst 42:20–39

Chapter 4
Fuzzy XML Queries and Index

Abstract Huge amounts of electronic data are available on the Internet, and XML has been the de-facto standard of information representation and exchange over the Web. The basic structure of XML is tree, and an XML query is often formed as a twig pattern with predicates additionally imposed on the contents or attribute values of the tree nodes. Also, the XML query technique based on index mechanism is developed to further improve the query efficiency. However, the XML fall short in their ability to handle imprecise and uncertain information in many real-world applications, and also the relevant XML query techniques cannot support twig pattern query in fuzzy XML. Currently, fuzzy XML data modeling has been extensively investigated as introduced in Chap. 3. Therefore, topics related to the querying of fuzzy XML can be considered very interesting in the fuzzy XML data context. In this chapter, we focus on the methods of fuzzy XML complex twig queries with predicates and of building index mechanism on fuzzy XML query.

4.1 Introduction

Information imprecision and uncertainty exist in many real-world applications and for this reason fuzzy XML data modeling has been extensively investigated as introduced in Chap. 3. Currently, as we have known, fruitful achievements about classic XML query have being done, and the XML query is usually formed as a twig pattern with predicates additionally imposed on the contents or attribute values of the tree nodes. However, the existing query methods cannot be applied to the fuzzy XML query directly. In order to meet the needs of practical application, it is not enough to only provide the modeling technology of fuzzy XML and fuzzy XML query is very necessary. Some efforts have been made in processing the incomplete and probabilistic XML data queries. Abiteboul and Senellart (2006) presented a model for representing probabilistic information in a semi-structured (XML) database, and studied the querying and updating probabilistic information in XML. Hung et al. (2003) proposed a model for probabilistic semi-structured data (PSD) and developed an extension of the relational algebra to handle

L. Yan et al., *Fuzzy XML Data Management*, Studies in Fuzziness and Soft Computing 311, DOI: 10.1007/978-3-642-44899-7_4,

probabilistic semi-structured data and described efficient algorithms for answering queries that use this algebra. Kimelfeld and Sagiv (2007) presented an algorithm for evaluating twig queries with projections over probabilistic XML documents. Other efforts on handling incomplete and probabilistic XML were done in Kimelfeld et al. (2008), Li et al. (2009) and Senellart and Abiteboul (2007). At present, the research about fuzzy twig query is relatively less. Liu et al. (2009) proposed a method to deal with fuzzy XML AND-logic query based on a twig pattern. Ma et al. (2011) presented the fuzzy query method of processing AND/OR logic with twig pattern, and this method is based on range encoding and uses the stack structure to store intermediate results. But, it should be noted that, in practical application, there are a lot of AND, OR and NOT logical predicates in the user's query conditions. Currently, only little research is being done about the fuzzy XML query including logic predicate twig, and the existing methods need a mass of computations, which may reduce the query efficiency because of the frequent I/O operations and the CPU consumption. Therefore, fuzzy XML query containing predicates is still an important topic in XML data management, and it is full of practical significance.

Moreover, in order to improve the efficiency of querying XML, many researchers suggest building index on the query algorithm. In the recent researches on XML twig pattern query, path index and sequence index are two methods of building index on XML data. The basic idea of path index is to transform XML documents into XML data graph, and then obtain path index figure by scanning XML data graph. Path index can be classified into three categories: classical path index, schema based path index and flattened structural path index. At present, the main classical path indexes of XML is DataGuide, 1-Index, D(k), and etc. (André et al. 2005; Chung et al. 2002). Sequence index is different and its main idea is to translate structure query into sequence matching. This can avoid time-consuming and complicated structural join operations in query. Wang et al. (2003) developed ViST, a dynamic indexing method for XML documents. The method supports structural XML queries by converting XML documents into sequences. The indexing method supports efficient non-contiguous sequence matching. A similar technique is used for weighted-subsequence matching and pattern discovery. The method also unifies structure indexes and value indexes into a single index that relies solely on B+Trees through a dynamic labeling method. The other efforts on building index on XML data can be found in Li and Moon (2001), Lu et al. (2011), Rao and Moon (2004) and Bruno et al. (2002). However, the research on fuzzy XML query based on the index mechanism is relatively less. Thus, building index for fuzzy XML data to improve the performance of query processing have received attention recently.

In this chapter, we focus on fuzzy XML queries and index. We present methods of fuzzy XML complex twig queries with AND, OR and NOT predicates, and investigate the problem of fuzzy XML query based on index.

4.2 Fuzzy XML Queries with Predicates

The fuzzy XML query containing logical predicates AND, OR and NOT is an important topic in the fuzzy XML data management. In this section, we present methods of fuzzy XML twig queries with AND, OR and NOT predicates. Before that, we first introduce fuzzy extended Dewey encoding, which is the basis of querying fuzzy XML data.

4.2.1 Fuzzy Extended Dewey Encoding

With the rapidly increasing popularity of XML for data representation, there is a lot of interest in query processing over data that conforms to a tree-structured data model. XML twig pattern matching is a key issue for XML query processing. XML documents can be modeled as ordered trees. Queries in XML query languages make use of twig patterns to match relevant portions of data in an XML database. The encoding of XML document tree is to assign a unique code to each node in the XML tree according to the certain rules. The purpose is to determine whether the two nodes are with ancestor/descendant, parent/child relationship or not, this process is done by the coding of any two nodes. So it can support the index and query of XML date more efficiently (Lu et al. 2005a, b).

At present, there are several types of encoding methods, such as range encoding and prefix encoding. Range encoding is a triple (*start*, *end*, *depth*), where "*start*" means the sequence code of the first visit to the node of the XML document tree in a depth-first traversal way, "*end*" means the sequence code of the second visit, and "*depth*" means the hierarchy value of the node in the XML document tree. Prefix encoding is known as the coding method based on path. An important prefix encoding way is the Dewey coding (Tatarinov et al. 2002). With Dewey Order, each node is assigned a vector that represents the path from the document's root to the node. Each component of the path represents the local order of an ancestor node. Dewey Order is "lossless" because each path uniquely identifies the absolute position of the node within the document. However, based on the original Dewey, we cannot know the element names along a path. To answer a twig query, we need to access the labels of all query nodes. Considering the fact that prefix comparison is less efficient than integer comparison, the performance of algorithm with the original Dewey is usually worse than that with region encoding. Further, Lu et al. (2005a) proposed an extension of Dewey encoding scheme, called extended Dewey. The extended Dewey label of each element can be efficiently generated by a depth-first traversal of the XML tree. Each extended Dewey label is considered as a vector of integers. We use label (*u*) to denote the extended Dewey label of element *u*. For each *u*, label (*u*) is defined as label (*s*):*x*, where *s* is the parent of *u*. The computation method of integer *x* in extended Dewey is a little more involved than that in the original Dewey. Assuming that $CT(t) = \{t_0, t_1, \ldots, t_{n-1}\}$ is the *child names clue* of tag *t*, for any element *u* with parent *s* in an XML tree, it follows:

1. If u is a text value, then $x = -1$;
2. Otherwise, assume that the element name of u is the kth tag in $CT(t_s)$ ($k = 0, 1, \ldots, n-1$), where, t_s denotes the tag of element s.
(a) If u is the first child of s, then $x = k$;
(b) Otherwise assume that the last component of the label of the left sibling of u is y (at this point, the left sibling of u has been labeled), then

$$x = \begin{cases} \left\lfloor \frac{y}{n} \right\rfloor \bullet n + k & if (y \bmod n) < k \\ \left\lceil \frac{y}{n} \right\rceil \bullet n + k & otherwise \end{cases}$$

Here n denotes the size of $CT(t_s)$.

The detailed introduction of the extended Dewey encoding can be found in Lu et al. (2005a). Note that the encoding methods mentioned above are just for classic XML document, and they cannot be directly applied to the fuzzy XML encoding. The encoding method cannot represent fuzzy information of fuzzy node, and also cannot effectively distinguish between vague nodes and precise nodes. Therefore, the coding scheme cannot support queries of fuzzy XML twig pattern. In the following, in order to support the encoding in the fuzzy XML document, we extend the method of the extended Dewey encoding. We add a new attribute called *FuzzySequence* on the basis of the original code, and it is an orderly collection that is used to store the membership degrees of the fuzzy nodes from the root node to the current node. That is, the fuzzy extended Dewey coding is a binary tuple (*Extended Dewey, FuzzySquence*). If there does not exist any fuzzy node from the root node to the current node, then we can say *FuzzySequence* is null. Figure 4.1 shows a practical application of the fuzzy extended Dewey coding, here we only intercept the first three layers of a fuzzy XML document.

It is shown in Fig. 4.1 that the fuzzy extended Dewey encoding still has a series of properties of the extended Dewey encoding (Lu et al. 2005a), and it further adds the membership degrees between the elements. Regarding the calculation of the

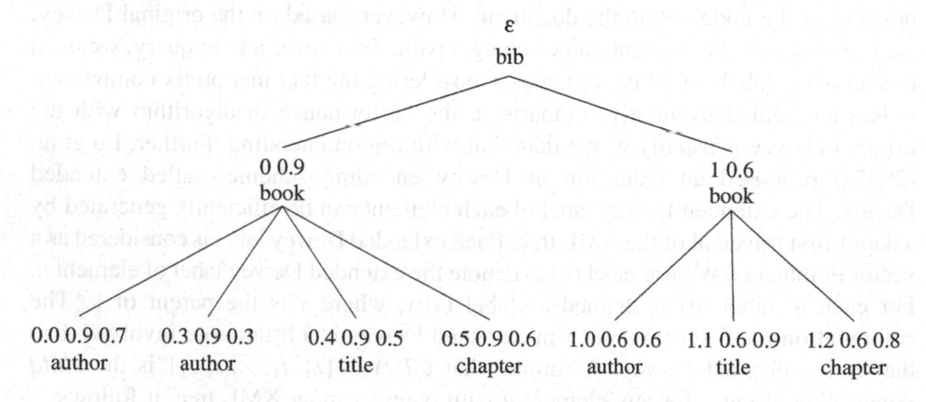

Fig. 4.1 The fuzzy extended Dewey coding of a fragment of fuzzy XML document

membership degrees, many of the existing approaches dealing with fuzzy information are based on the theory of fuzzy sets (Zadeh 1965). The Zadeh's minimum and maximum operations can be used to calculate the intersection and union of the memberships. However, in the process of applying to fuzzy XML twig query, the results of fuzzy intersection and fuzzy union obtained by Zadeh's method is unsuitable.

Let us look at an example. Suppose that there is the following membership information of a path: $\delta\ (a \rightarrow b) = 0.4$, $\delta\ (a \rightarrow c) = 0.7$, $\delta\ (a \rightarrow d) = 0.8$. Then when we use the Zadeh's method to calculate it, we can get $\delta\ (a \rightarrow b, a \rightarrow c) = \delta\ (a \rightarrow b, a \rightarrow d) = 0.4$. But, in practice, two results above should be different. Also, if there are conditions: $\delta\ (a \rightarrow b, a \rightarrow c) = 0.4$ and $\delta\ (a \rightarrow b) = 0.4$, then by the reverse operation of the Zadeh's method, we cannot obtain the specific value of $\delta\ (a \rightarrow c)$, and only can get a ballpark range.

Therefore, here we use the product of *t*-norm instead of Zadeh's logical product. In this case, the intersection of the membership degrees is calculated as follows: $\delta_{\mathrm{i} \cap \mathrm{j}} = \delta_{\mathrm{i}} \times \delta_{\mathrm{j}}$. We use δ_{k} $(k = 1, 2,\ldots, n)$ to denote the relative membership degree of a fuzzy XML fragment *F* and use δ_{whole} to denote the absolute membership degree of *F*. Then we have $\delta_{\mathrm{whole}} = \Pi_{k=1,\ldots,n}\delta_{\mathrm{k}}$. Moreover, suppose $\delta\ (Q)$ is the threshold for a given user query. Then we have

1. if $\delta_{\mathrm{whole}} \geq \delta\ (Q)$, then $\delta_{\mathrm{k}} \geq \delta\ (Q)$;
2. if $\delta_{\mathrm{k}} \geq \delta\ (Q)$, then $\Pi_{t=1,\ldots,m}\delta_{\mathrm{t}} = \delta_{\mathrm{whole}} / \Pi_{t=m+1,\ldots,n}\delta_{\mathrm{t}}$.

In addition, because in the process of the OR-logical twig query, as long as one of the OR branches meets the query condition, the query will be executed. At this point, the maximum operation of Zadeh's method can satisfy the demand well, and thus the union of membership degrees can be defined as follows: $\delta_{\mathrm{i} \cup \mathrm{j}} = \max\ (\delta_{\mathrm{i}}, \delta_{\mathrm{j}})$. Based on the fuzzy extended Dewey encoding, in the following we introduce fuzzy XML twig queries with AND, OR and NOT predicates.

4.2.2 Fuzzy XML Twig Query with AND-Logic

Regarding the twig query *Q*, we define several node functions as follows: *IsLeaf* (*n*): *Node* → *Bool*, determining whether the node *n* is a leaf node; *IsBranching* (*n*): *Node* → *Bool*, determining whether the node *n* is the branch node; *LeafNodes* (*n*): *tNode* → {*Node*}, returning a collection of leaf nodes whose roots are the node *n* in twig pattern query *Q*; *dbl* (*n*): *Node* → {*Node*}, returning the branch node *b* and leaf node *f* satisfying the conditions: (1) the roots of *b* and *f* are the node *n* in twig pattern query *Q*; (2) there is no any node from the path *n* to *b* or *f*. Moreover, in fuzzy XML document tree, there is a list T_{f} associated with each leaf node *f*, which includes the fuzzy extended Dewey coding of the elements whose labels are *f*. Several operations can be executed to the list T_{f}, such as *current* (T_{f}), *advance* (T_{f})

and *eof* (T_f). The function *current* (T_f) returns the fuzzy extended Dewey coding of the current element. We can obtain the fuzzy extended Dewey coding of the current element through *current* (T_f).*number*, and obtain the path information of the current element through *current* (T_f).*fs*. The function *advance* (T_f) points to a new current element which is the next element of the current element. The function *eof* (T_f) is to test whether it is the footer. For an element *e* in fuzzy XML documents, there are two operation functions: *ancestors* (*e*) and *descendants* (*e*), which return the ancestor elements and the descendant elements of *e* respectively. In addition, the stack structure is also used in the algorithm. Each branch node *b* associates with a collection of S_b, and any a pair of elements in S_b is either the ancestor/descendant relationship or the parent/child relationship.

Now we introduce an algorithm of fuzzy XML twig query with AND-logic, called FATJFast, which is the extension of the existing TJFast algorithm (Lu et al. 2005a, b). The algorithm is a query matching method based on path. Given the fuzzy extended Dewey coding of a node, according to the characteristic of the visibility of its ancestor' name, it is easy to decide whether the single path from the root node to the node meets the path corresponding to the twig query. Therefore, the key problem of query is the integration between the single path and the other single paths of meeting the query conditions, and after the integration we can obtain the final twig matching results. Moreover, a filter is applied in the algorithm: the first layer is to decide whether the fuzzy information of a single path meets the requirement of the user's query threshold; the second filter is to check whether the whole membership degree of the integrated matching results meets the requirement of the user's query threshold.

Definition 4.1 (*model of query path P_f*) Given a query node *f* of the query twig, the names of the query nodes from the root node to the node *f* and the structure relationships of the query nodes are called a model of the query path of *f*, marked as P_f.

Definition 4.2 (*associated element*) Given a query node *f* in the query twig and an element *e* in a fuzzy XML document, if their labels are consistent, namely, *Label* (*f*) = *Label* (*e*), and the element *e* meets the query path model P_f, we say that *e* is the associated element of *f*.

Definition 4.3 (*subquery* (n_i, n_j)) Given a sub-path query subquery (n_i, n_j) and an element e_i in a fuzzy XML document *D*, we say that e_i satisfies subquery (n_i, n_j), if

1. the element e_i has the same label name with the node n_i;
2. one of the following two conditions holds: (a) $i = j$; or (b) *edge* (n_i, n_{i+1}) is an ancestor/descendant relationship (or parent/child relationship) edge, and there exists an element e_{i+1} in *D* such that the *edge*(e_i, e_{i+1}) is ancestor/descendant relationship (or parent/child relationship) and the element e_{i+1} satisfies subquery (n_{i+1}, n_j); and
3. the membership degree of the path meets the requirement of the user's query threshold, i.e., $\delta_{ei\rightarrow ej} \geq \delta_Q$.

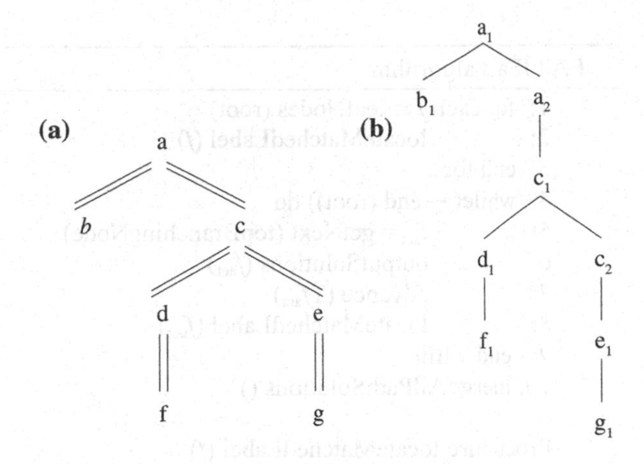

Fig. 4.2 Twig query and XML document

Definition 4.4 Given a twig query Q and a fuzzy XML document D, e is the associated element of the branch node b in Q. For $\forall n_i \in dbl\ (n)$ and its corresponding associated elements e_i, if the following conditions hold, then we say that e has the expansion of the solution w.r.t. the branch node b:

1. n_i is a leaf node and e_i meets twig query subquery (n, n_i); or
2. n_i is a branch node, e_i meets twig query subquery (n, n_i), and e_i has the expansion of the solution w.r.t. n_i.

The twig query without branch nodes is called path query. If the associated element meets the sub-path twig query of its corresponding node, then we say that it has the expansion of the solution. In our FATJFast algorithm, we always check the elements in the flow of leaf nodes, calculate and determine the membership degrees of their corresponding paths in the process of matching, and determine the whole membership degree of the matching results in the integration stage of the paths.

For example, given a query twig in (a) and an XML document in (b) of Fig. 4.2 (for simplicity, the fuzzy information in the XML document is omitted here), when the inspection elements of the query nodes are a_1, b_1, c_1, d_1, e_1, f_1, and g_1, since b, $c \in dbl\ (a)$ and b is the leaf node, the b_1 meets the sub-path query; Next, we further check the twig nodes $c, f, g \in dbl\ (c)$, since f_1 and e_1 meet the sub-path queries subquery (c, f) and subquery (c, g) respectively, the element c_1 has the expansion of the solution w.r.t. the twig node c. Based on the observation above, the element a_1 has the expansion of the solution w.r.t. the branch node a. When the inspection elements are a_1, b_1, c_2, d_1, e_1, f_1, and g_1, the element c_2 does not have the expansion of the solution w.r.t. the branch node c.

The following is the FATJFast algorithm. The algorithm includes two main stages: the first stage (lines 1–9) finds the path matching results of querying leaf nodes; and the second stage (line 10) gets the final match results by integrating all the single path results.

FATJFast algorithm

```
1: for each f ∈ leafNodes (root)
2:         locateMatchedLabel (f)
3: end for
4: while(¬end (root)) do
5:         f_act = getNext (topBranchingNode)
6:         outputSolutions (f_act)
7:         advance (Tf_act)
8:         locateMatchedLabel (f_act)
9: end while
10: mergeAllPathSolutions ()
```

Procedure locateMatchedLabel (*f*)

1: while $\neg((n_1/n_2/\ldots/n_k$ matches pattern $p_f) \wedge (n_k$ matches $f) \wedge (\delta_{pf} \geq u))$ do
2: advance (T_f)
3: end while

Function end (*n*)

1: Return $\forall f \in$ leafNodes $(n) \rightarrow$ eof (T_f)

Procedure outputSolution (*f*)

1: Output the path solutions of the current (T_f) to pattern p_f such that in each solution s, $\forall e \in s$, the element e matches a branching node $b \rightarrow e \in S_b$.

Procedure mergeAllPathSolutions ()

```
1: for each Q
2:         if(δ_whole (Q) ≥ δ)
3:                 for each path answer p_i
4:                         add p_i to the corresponding list
5:                 end for
6:         end if
7: end for
8: extend all the answers in the topBranchingNode's list to the final answers
```

In lines 1–3, in each stream of leaf nodes, the function *locateMatchedLabel* can find the first element which satisfies the path pattern of the leaf nodes. In line 5, we identify the next stream T_{fact}, which will be processed by invoking *getNext* (*topBranchingNode*), where *topBranchingNode* is the branch node at the highest level, i.e., the ancestor of all other branch nodes. In line 6, we output some path matching solutions, in which each element is in the corresponding branch node set S_b. In lines 7 and 8, the function *advance* (T_{fact}) can locate the next matching element, and then the procedures in lines 1–3 are repeated. In the procedure of *locateMatchedLabel*, the labels and structures of the single path elements will be matched based on the feature of the visibility of the ancestors' names in fuzzy extended Dewey coding. Also the query threshold of the single path should be satisfied, i.e., $\delta_{pf} \geq u$, where δ_{pf} denotes the membership degree of the path solutions of the leaf node *f*. Note that, δ_{pf} is the product of the membership degrees of all the elements in query paths, and the membership degree of each element can be obtained from the attribute *FuzzySequence* in the fuzzy extended Dewey coding. In *locateMatchedLabel*, according to the requirement of the user's threshold, we can quickly skip the elements that cannot form the final solution, and thus avoid a lot of unnecessary operations. For the procedure of *mergeAllPathSolutions*, inspired by the idea of the blocking techniques in Lu et al. (2005b), we create two lists associated with each element *n* in the sets of branch notes, i.e., *(S)elf-list*, which stores the blocked solutions whose root elements are *n*, and *(I)nherit-list*, which stores the blocked solutions whose root elements are the descendant nodes of *n*. At any point of the algorithm, we do not directly output the path solution of an element, but add it into the *Self-list* of the nearest branching node corresponding to the element. Moreover, the core operation *getNext* () in line 5 in the FATJFast algorithm is as follows.

Function *getNext* () in the FATJFast algorithm

```
1: if (isLeaf (n)) then
2: return n
3: else
4: for each n_i ∈ dbl (n) do
5:         f_i = getNext (n_i)
6:         if (isBranching (n_i) ∧ empty (Sn_i))
7:                 return f_i
8:         e_i = max {p|p ∈ MB (n_i, n)}
9: end for
10:        max = maxarg_i {e_i}
11:        min = minarg {e_i}
12: for each n_i ∈ dbl (n) do
13:        if (∀ e ∈ MB(n_i, n):e ∉ ancestors (e_max))
14:                return f_i
15:        end if
16: end for
17: for each e ∈ MB (n_min, n)
18:        if (e ∈ ancestors (e_max)) updateSet (Sn, e)
19: end for
20:        return f_min
21: end else
22: end if
```

Function MB (n, b)
1: if (isBranching (n)) then
2: let e be the maximal element in set Sn
3: else
4: let e = current (Tn)
5: end if
6: Return a set of element a that is an ancestor of e such that a can match node b in the path solution of e to path pattern p_n

Procedure clearest (S, e)
1: Delete any element a in the set S such that $a \notin$ ancestors (e) and $a \notin$ des cendants (e)

Procedure updateSet (S, e)
1: clearest(S, e)
2: Add e to set S

The function *getNext* is the core of the FATJFast algorithm, and it includes two main parts. For the first part, the function getNext (n) is to identify the next file stream to be processed, and return the query leaf node f according to the following recursive procedures:

1. if n is a leaf node, then n is returned (line 2);
2. if n is a branch node, then for each node $n_i \in dbl$ (n), the following three operations are executed:

   (a) if the current element in the file stream T_{fi} does not match the sub-tree whose root is the branch node n_i, i.e., there is not the extension of the solution, f_i is returned (line 7);
   (b) if the current element in the file stream T_{fi} does not merge with the elements in the other file streams to form the final matching result, f_i is returned (line 14); otherwise,
   (c) if the element has the extension of the solution, the node f_{min} is returned, in which the element e_{min} corresponding to f_{min} has the minimal coding by lexicographical order (line 20).

Moreover, the function MB (n_i, n) (where $n_i \in dbl$ (n)) is to find the set of elements that meet the subquery (n, n_i), and it is an important auxiliary function to judge whether two paths can integrate to form the final matching result. The second part of the function *getNext* is to update the set S_b associated with the branch node b.

In the following, we provide an example to illustrate the FATJFast algorithm. Given a query twig as shown in Fig. 4.3a and the corresponding fuzzy XML document in (b). In the fuzzy XML document, there are four branch structures conforming to the twig query, and each branch has the different membership degrees. Assume that the query threshold given by the user is 0.2, the first branch meets the structure requirement and the membership degree also meets the user's threshold requirement. The second branch meets the structure requirement, but the membership degree does not meet the user's requirement. The membership degree in the third one does not meet the requirement of the threshold. The structure in the forth one does not meet the first layer of the filter.

In Fig. 4.3, there are two input lists, T_c and T_d. First, for the node *c*, the element c_1 meets the structure matching of the path, and the membership degree $\delta_{pc} = 0.9 \times 0.8 = 0.72$ meets the requirement of threshold. Thus the element c_1 is positioned in the file stream of *c*. Similarly, the element d_1 also meets the requirements of the structure and the threshold. The function getNext (*b*) recursively calls getNext (*c*) and getNext (*d*), and *c* and *d* are leaf nodes. So getNext (*c*) = *c* and getNext (*d*) = *d*. Then, getNext (*b*) is executed and we have MB (*c*, *b*) = $\{b_1\}$, MB (*d*, *b*) = $\{b_1\}$, and $e_{min} = e_{max} = b_1$. Next, b_1 is inserted into the stack S_b corresponding to the node *b*. When b_1 is inserted into the stack S_b, the whole membership degree of the single paths $<b_1, c_1>$ and $<b_1, d_1>$ will be calculated. If the whole membership degree $\delta_{whole}(b_1)$ meets the requirement of the threshold, the block technology will be used, and the paths $<b_1, c_1>$ and $<b_1, d_1>$ will be pressured into the list *Self-list* of b_1 and considered as the final output. Further, the node *c* corresponding to the smallest coding element node c_1 is retuned. In this case, the pointer of the file stream T_c points to the next element c_2 (line 7) and then the path is matched. Since the structure matching path is $(a \to b \to c)$ and the path membership degree $\delta_{pc} = 0.4$ meets the requirement of user's threshold, the algorithm goes into the second round of getNext processing.

The process of the second round is similar to the first round above. First, MB (*c*, *b*) = $\{b_2\}$, MB (*d*, *b*) = $\{b_1\}$, $e_{min} = b_1$, $e_{max} = b_2$, and $e_{min} \notin$ ancestors (e_{max}), therefore, the leaf node *d* is returned. Then, the file stream T_d points to the next element d_2, and the algorithm turns into the third round of *getNext*. Next, the

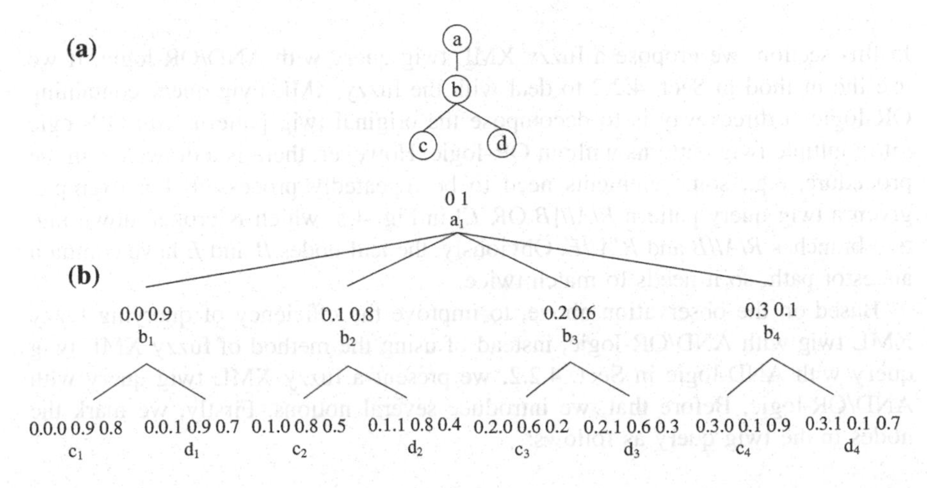

Fig. 4.3 A query example of the FATJFast algorithm. **a** A twig query. **b** A fuzzy XML document

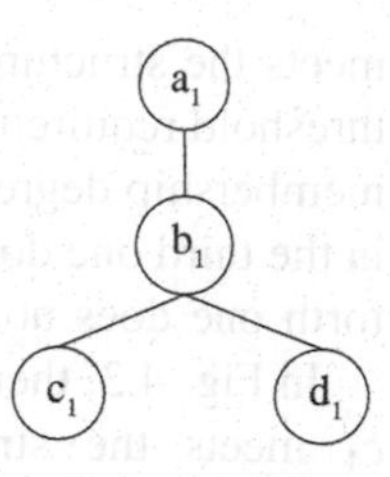

Fig. 4.4 The final matching result of the query in Fig. 4.3

element b_2 is pressed into the stack and b_1 is popped from the stack. Note that the whole membership degree of the paths $<b_2, c_2>$ and $<b_2, d_2>$ does not meet the requirement of the threshold, so the $<b_2, c_2>$ and $<b_2, d_2>$ will not pressured into the list *Self-list* of b_2. Finally, all of the file streams are handled and we get the final matching result (a_1, b_1, c_1, d_1) as shown in Fig. 4.4.

Note that, due to the visible feature of the ancestors in fuzzy extended Dewey encoding, our algorithm FATJFast only needs to access the data streams corresponding to the leaf nodes instead of accessing all of the data streams of the query nodes, which may reduce the frequency of the I/O operations and improve the efficiency of querying. Regarding the complexity of the algorithm, for a fuzzy XML document *D* and a twig query *Q*, there are only the ancestor/descendant relationships between the branch nodes and its son nodes in *Q*. At worst, the I/O complexity has the linear relationship with the size of input and output lists. The worst-case space complexity is $O(d^2 * |b| + d * |f|)$, here $|f|$ denotes the number of leaf nodes, $|b|$ the number of the branch nodes in *Q*, and *d* the length of the coding of the longest element in the input list.

4.2.3 Fuzzy XML Twig Query with AND/OR-Logic

In this section, we propose a fuzzy XML twig query with AND/OR-logic. If we use the method in Sect. 4.2.2 to deal with the fuzzy XML twig query containing OR-logic, a direct way is to decompose the original twig pattern with OR-logic into multiple twig patterns without OR-logic. However, there is a drawback in the procedure, e.g., some elements need to be repeatedly processed. For example, given a twig query pattern *R/A//[B* OR *E]* in Fig. 4.5, which is broken down into two branches *R/A//B* and *R/A//E*. Obviously, the leaf nodes *B* and *E* have common ancestor path, so it needs to match twice.

Based on the observation above, to improve the efficiency of querying fuzzy XML twig with AND/OR-logic, instead of using the method of fuzzy XML twig query with AND-logic in Sect. 4.2.2, we present a fuzzy XML twig query with AND/OR-logic. Before that, we introduce several notions. Firstly, we mark the nodes in the twig query as follows:

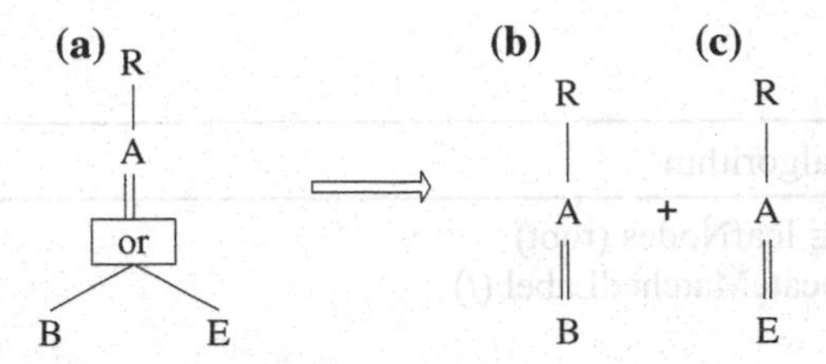

Fig. 4.5 Transformation from a twig pattern with OR-logic to the twig patterns with AND-logic

- the AND-logic nodes in the query tree are marked as *Anode*;
- the OR-logic nodes in the query tree are marked as *ONode*;
- the other query node is marked as *QNode*.

Then, the following specific principles in the twig query are defined:

- if a *Anode* node or *ONode* node *n* has a child node n_i with the same type, then n_i can be removed and all child nodes of n_i are connected to the node *n* as the child nodes of *n* as shown from (c) to (d) in Fig. 4.6.
- if a query node *QNode n* has the *Anode* node n_i as its child node, then n_i can be removed and all child nodes of n_i are considered as the child nodes of *n* as shown from (a) to (b) in Fig. 4.6.

Moreover, we define a function *qNode*(*n*): *Node* → *Node*, if *n* is *QNode*, it returns *Label*(*n*); otherwise, *n* is a predicate node, it returns the tag of the parent node of *n*.

In the following, we give a fuzzy XML twig query algorithm with AND/OR-logic, called FA/OTJFast. In the algorithm FA/OTJFast, the function *locate-MatchedLabel* () is still used to filter the leaf elements as mentioned in the FATJFast algorithm in Sect. 4.2.2.

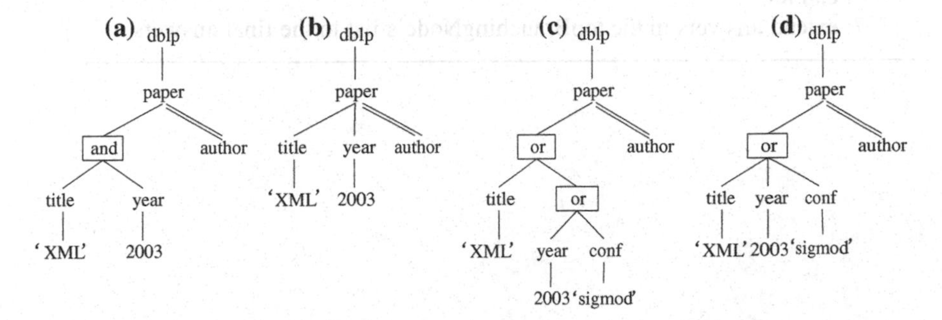

Fig. 4.6 Normalization of the twig query with AND/OR-logic predicates

FA/OTJFast algorithm

```
1: for each f ∈ leafNodes (root)
2:         locateMatchedLabel (f)
3: end for
4: while (¬end (root)) do
5:         f_act = A/OgetNext (topBranchingNode)
6:         outputSolutions (f_act)
7:         advance (T_fact)
8:         locateMatchedLabel (f_act)
9: end while
10: mergeAllPathSolutions ()
```

Procedure locateMatchedLabel (*f*)

1: while $\neg((n_1/\ldots/n_k$ matches pattern $p_f) \wedge (n_k$ matches $f) \wedge (\delta_{pf} \geq u))$ do
2: advance (T_f)

Function end (*n*)

1: Return $\forall f \in$ leafNodes $(n) \rightarrow$ eof (T_f)

Procedure outputSolution (*f*)

1: Output the path solutions of the current (T_f) to pattern p_f such that in each solution *s*, $\forall e \in s$, the element *e* matches a branching node $b \rightarrow e \in S_b$.

Procedure mergeAllPathSolutions ()

```
1: for each Q
2: if (δ_whole (Q) ≥ δ)
3:         for each path answer p_i
4:                 add p_i to the corresponding list
5:         end for
6: end for
7: extend answers in the topBranchingNode's list to the final answers
```

The core function A/OgetNext () in the FA/OTJFast algorithm is as follows:

A/OgetNext ()

```
1: if (isLeaf (n)) then
2:         return n
3: else
4:         for each n_i ∈ dbl (n) do
5:         f_i = getNext (n_i)
6:         if (isBranching (n_i) ∧ empty (Sn_i))
7:               return f_i
8:         e_i = max{p | p ∈ MB (n_i, n)}
9:         end for
10:        if (n is an ONode)
11:        for each e ∈ MB (n_min, n) do updateStack(Sn, e)
12:              min = minarg {e_i}
13:              for each n_i ∈ dbl (n) do
14:              if(∀ e ∈ MB (n_min, n), ∀ e_i ∈ MB (n_i, n): e ∈ ancestors (e_i))
15:                                  δ_i = δ_n→ni
16:              max = maxarg_i {δ_i}
17:              return f_max
18:        end for
19:        else
20:              max = maxarg_i {e_i}
21:              min = minarg {e_i}
22:              for each n_i ∈ dbl (n) do
23:                        if (∀ e ∈ MB (n_i, n): e ∉ ancestors (e_max))
24:                              return f_i
25:                        end if
26:              end for
27:              for each e ∈ MB (n_min, n)
28:                        if (e ∈ ancestors (e_max)) updateSet (Sn, e)
29:              end for
30:              return f_min
31: end else
32: end if
```

Function MB (n, b)
1: if (qNode (b) == qNode (n)) then return the set of elements in set Sn
2: if (isBranching (n)) then
3: Let e be the maximal element in set Sn
4: else
5: Let e = current (Tn)
6: Return a that is an ancestor of e , such that a can match qNode (b) in the path solution of e to path pattern p_n

Procedure clearSet (S, e)
1: Delete any element a in set S such that $a \notin$ ancestors (e) and $a \notin$ descendants (e)

Procedure updateSet (S, e)
1: if (e is in S)
2: return
3: else
4: clearSet (S, e)
5: Add qNode (e) to set S

The function A/OgetNext is an extension of the function getNext () mentioned in the algorithm FATJFast in Sect. 4.2.2. The function A/OgetNext is still to complete two tasks: one is to determine the next pending file stream; and another is to update the stack S_b corresponding to the branch node b. For the first part, the function A/OgetNext (n) is to identify the next file stream to be processed, and returns query leaf nodes f according to the following recursive procedures:

1. if n is a leaf node, then n is returned (line 2);
2. if n is a branch node, then for each node $n_i \in dbl$ (n), the following four operations are executed:
   (a) if the current element in the file stream T_{fi} does not match the sub-tree whose root is the branch node n_i, f_i is returned (line 7);
   (b) if n is a *ONode* node, then there is a node which has the smallest coding and the expansion of the solution w.r.t. *qNode* (n). In this case, in all of the branches of the node, the leaf node f_{max} which has the largest membership degree is returned (line 17);
   (c) if n is a *ANode* or *QNode* node, and the current element in the file stream T_{fi} does not merge with the elements in the other file streams to form the final matching result, f_i is returned (line 24); otherwise,
   (d) the node f_{min} is returned, here the element e_{min} corresponding to f_{min} has the minimal coding by lexicographical order (line 30).

Moreover, the function MB (n_i, n) (where $n_i \in dbl$ (n)) in the FATJFast algorithm is further extended. In the extended MB (n_i, n), according to the associated elements of the query node n_i, it can find a set of elements of matching with the node

n in the query path P_{ni}. Due to the specificity of the OR predicate nodes, when n_i is a *ONode* node, i.e., *qNode* (n_i) = n (n is the direct branch node of n_i), the extended function MB (n_i, n) directly returns the elements in the branch node stack *Sn*.

In the following, we give an example to illustrate the FA/OTJFast algorithm. Given a query twig with AND/OR-logic as shown in Fig. 4.7a and a fuzzy XML document in (b). Assume that the query threshold given by the user is 0.1. At first, the function A/OgetNext (a) recursively calls A/OgetNext (OR) and A/OgetNext (e) [where OR , $e \in dbl$ (a)]. Since e is a leaf node, it returns A/OgetNext (e) = e; OR is a branch node, A/OgetNext (OR) recursively calls A/OgetNext (b) and A/OgetNext (d), it returns A/OgetNext (b) = b, A/OgetNext (d) = d. Then, the algorithm calculates MB (n_{min}, n) = a_1, presses the element a_1 into the stack S_a, inserts the path result ($a_1 \rightarrow b_1$) into the list *Self-list* of the branch node a, and returns the leaf node under the OR branch which has the largest membership degree, i.e., A/OgetNext (OR) = b. Next, the algorithm continues to getNext (a) and computes MB (OR, a) = $\{a_1\}$ and MB $\{e, b\}$ = $\{a_1\}$. After that, the algorithm inserts a_1 into the stack S_a, pressures the path structure ($a_1 \rightarrow e_1$) into the list *Self-list* of a, and calculates the whole membership degree $\delta_{whole} = 0.504$, which meets the threshold requirement. Finally, we get the final solutions ($a_1 \rightarrow e_1$) and ($a_3 \rightarrow e_3$) as shown in Fig. 4.8.

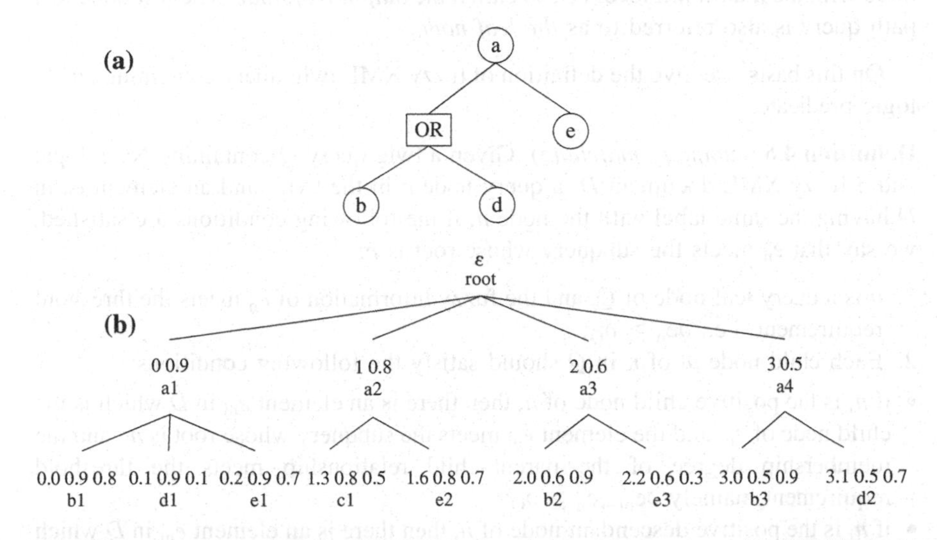

Fig. 4.7 A query example of the FA/OTJFast algorithm. **a** A twig query with AND/OR logic predicates. **b** A fuzzy XML document

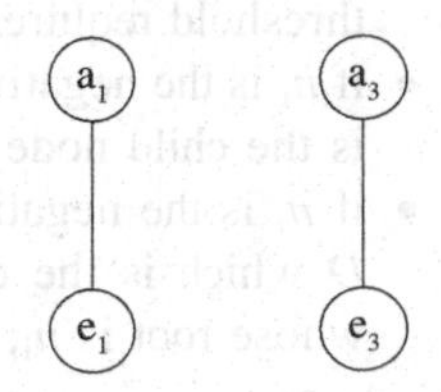

Fig. 4.8 The final query matching results

4.2.4 Fuzzy XML Twig Query with NOT-Logic

In this section, we extend the Ntjfsat¬ algorithm (Garakani et al. 2007) to resolve the fuzzy XML twig query containing NOT-logic predicate, and propose a fuzzy XML twig query with NOT-Logic algorithm, called FNTJFast.

Every twig query has the corresponding query nodes and the structural relationships among query nodes. A twig query Q containing NOT-logical predicate is a set of query nodes $<n_1, n_2, \ldots, n_m>$, the node n_j is connected with its ancestor node by edge $edge(n_i, n_j)$. Due to the specificity of the NOT predicate, the relationships among nodes are divided into the following four types:

- the positive parent–child relationships, connected by single lines '|';
- the positive ancestor–descendant relationships, connected by double lines '||';
- the negative parent–child relationships, connected by '¬|';
- the negative ancestor–descendant relationships, connected by '¬||'.

Definition 4.5 (*output node, non-output node, output leaf node, leaf node*) A node n_i in a path query is regarded as an *output node* if there is no any negative edges from the root node to the node n_i; otherwise, it is a *non-output node*. The output node with the maximum level is also called the *output leaf node*. The last node in a path query is also referred to as *the leaf node*.

On this basis, we give the definition of fuzzy XML twig query containing NOT-logic predicate.

Definition 4.6 (*subquery matching*) Given a twig query Q containing NOT-logic and a fuzzy XML document D, a query node n in the twig, and an element e_n in D having the same label with the node n, if the following conditions are satisfied, we say that e_n meets the subquery whose root is n:

1. n is a query leaf node of Q, and the fuzzy information of e_n meets the threshold requirement, i.e., $\delta e_n \geq \delta_Q$;
2. Each child node n_i of n in Q should satisfy the following conditions:

- if n_i is the positive child node of n, then there is an element e_{ni} in D which is the child node of e_n, and the element e_{ni} meets the subquery whose root is n_i, and the membership degree of the parent–child relationship meets the threshold requirement, namely $\delta e_{ni \to} e_n \geq \delta_Q$;
- if n_i is the positive descendant node of n, then there is an element e_{ni} in D which is the descendant node of e_n, and the element e_{ni} meets the subquery whose root is n_i, and the membership degree of the parent–child relationship meets the threshold requirement, namely $\delta e_{ni \to} e_n \geq \delta_Q$;
- if n_i is the negative child node of n, then there is no any element e_{ni} in D which is the child node of e_n, and the node e_{ni} meets the subquery whose root is n_i;
- if n_i is the negative descendant node of n, then there is no any element e_{ni} in D which is the descendant node of e_n, and the node e_{ni} meets the subquery whose root is n_i;

3. For the matching results of meeting the conditions of (1) and (2), the whole membership degree δ_{whole} meets the threshold requirement, namely $\delta_{\text{whole}} \geq \delta_{\text{Q}}$.

Definition 4.7 (*output element of NOT-logic twig query*) Given a twig query Q and a fuzzy XML document D, there are k output nodes $<n_1, n_2,\ldots, n_k>$ in Q, if a set of elements $<e_1, e_2,\ldots, e_k>$ in D satisfies the following conditions, we say that $<e_1, e_2,\ldots, e_k>$ is the output element of the final matching query result:

1. the element e_i and query node n_i have the same tag name;
2. each output element e_k should meet the subquery whose root is n_k.

Here we take Fig. 4.9 as an example to illustrate the definitions above, and for simplify we don't consider the membership degrees of the elements.

In Fig. 4.9, there is a descendant element C_1 under B_1 such that C_1 meets the subquery *C//D* whose root is *C*, but B_1 does not meet the subquery whose root is *B*, so $<A_1, B_1>$ is not the output element, and only $<A_1, B_2>$ is the output element.

In the following, we give the fuzzy XML twig query with NOT-logic, called FNTJFast. In the process of matching, the algorithm only accesses the *positive leaf nodes* that may be the final match results and the *negative leaf nodes* that can help to filter the final results. Compared with the existed twig matching algorithms, the algorithm FNTJFast may effectively shorten the matching time and reduce the number of accessing elements. The algorithm first extracts the leaf nodes (including the positive leaf nodes and the negative leaf nodes) in the query twig Q, and then according to the finite state transducer FST of fuzzy XML documents (Lu et al. 2005a, b), finds the file streams in all levels that have the same tag names with the leaf nodes and simultaneously filters some file streams which cannot merge with the other path results to form the final positive matching results. Finally, as the requirement of the membership degree, the algorithm returns the final matching results.

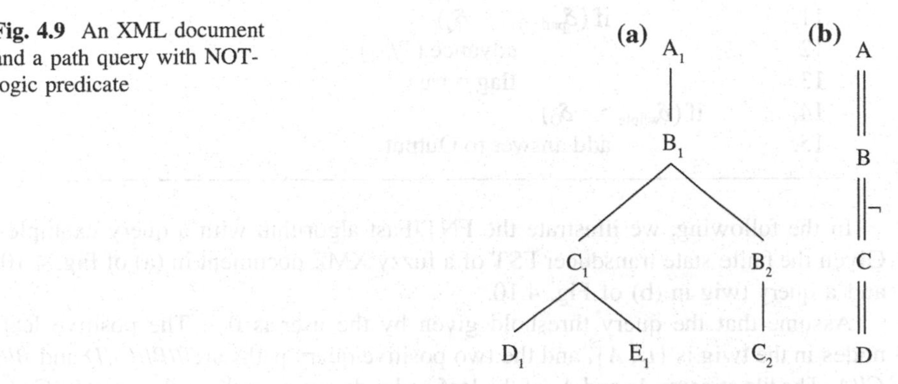

Fig. 4.9 An XML document and a path query with NOT-logic predicate

```
The FNTJFast algorithm
1.   while (¬end (root))
2.        for each fi ∈ Pos_leafNodes (root)
3.             Locate Tfi
4.        for each fi ∈ Pos_leafNodes (root)
5.             locateMatchedLevel (fi, fi+1)
6.        if (δwhole >= δQ)
7.             add answer to Output
8.        if (there is neg_extention)
9.             for each nfi ∈ Neg_leafNodes (root)
10.                 for each element nf in Tnfi
11.                 for each answer in Output
12.                 if (n.prefix (Lnfi, ans) == answer.prefix (Lnfi, ans))
13.                      remove the answer from Output

Procedure Locate Tfi
1.   locate Tfi for leaf node fi that can merge with other paths in the twig

Function locateMatchedLevel (fi, fi+1)
1.   flag = true
2.   while (flag)
3.        if (current (Tfi).number.prefix (Li, i+1) <
               current (Tfi+1).number.prefix (Li, i+1))
4.             advance (Tfi)
5.        else if (current (Tfi).number.prefix (Li, i+1) >
               current (Tfi+1).number.prefix (Li, i+1))
6.             advance (Tfi+1)
7.        else
8.             if (δpath_fi < δQ)
9.                  advance(Tfi)
10.                 flag = true
11.            if (δpath_fi+1 < δQ)
12.                 advance (Tfi+1)
13.                 flag = true
14.       if (δwhole >= δQ)
15.            add answer to Output
```

In the following, we illustrate the FNTJFast algorithm with a query example. Given the finite state transducer FST of a fuzzy XML document in (a) of Fig. 4.10 and a query twig in (b) of Fig. 4.10.

Assume that the query threshold given by the user is 0.3. The positive leaf nodes in the twig is $\{D, A\}$, and the two positive query paths are *//B//C//D* and *B//C//A*. The file streams A_1 and A_2 of the leaf nodes do not meet the path query *B//C//A*.

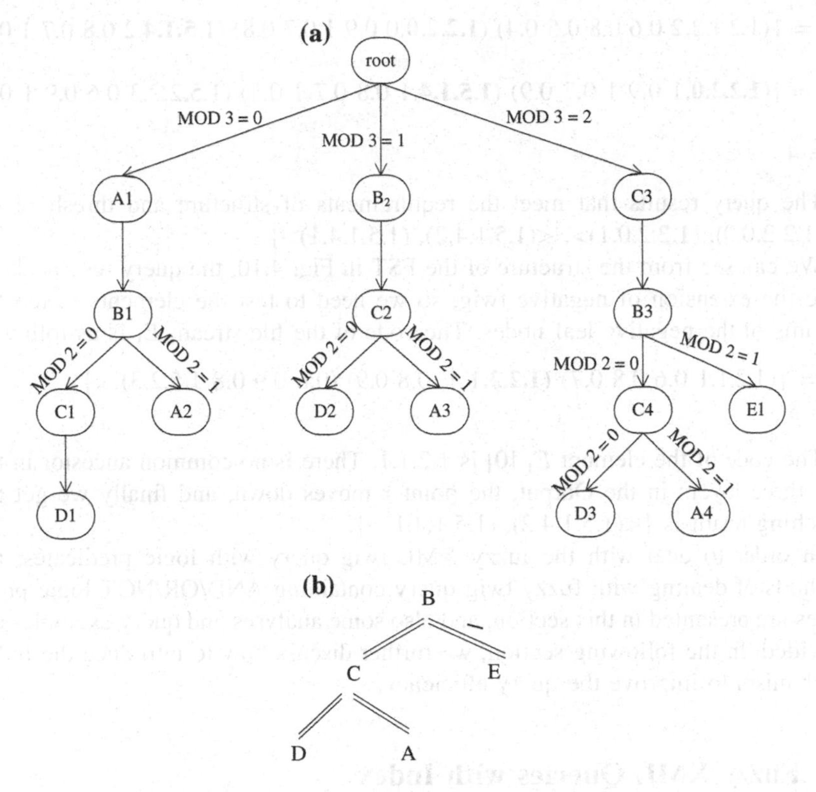

Fig. 4.10 A query example of the FNTJFast algorithm. **a** The finite state transducer FST of a fuzzy XML document. **b** A query twig with NOT-logic

Although the file stream D_1 of the leaf node D meets the path query *//B//C//D*, it cannot integrate in the twig node C with the file stream of the node A to form the matching results, so the algorithm skips the file streams A_1, A_2 and D_1 in the process of Locating. Moreover, the file streams D_2 and A_3 not only meet the path query but also can integrate to form the final match results. The codes of D_2 and A_3 are as follows:

D2 = {(1.1.1.0 0.8 0.6 0.2) (1.4.1.2 0.8 0.7 0.9) (**1.4.2**.4 0.8 0.9 0.8) …}
A3 = {(1.1.2.1 0.8 0.6 0.6) (1.4.2.3 0.8 0.9 0.8) (**1.4.2**.5 0.7 0.6 0.5) …}
L = 3

We further examine the elements in D_2 and A_3 one by one. The third element in D_2 is $D_2[2]$, which has the same code with $A_3[1]$ in the first three layers. Then, we calculate the membership degrees of their paths: 0.576 and 0.576, and we can see that the two degrees are bigger than the query threshold 0.3. Further, we calculate the whole membership degree $\delta_{\text{whole}} = (0.8 \times 0.9) \times 0.8 \times 0.8 = 0.4608$, which meets the threshold requirement. Therefore, we store the extension of Dewey coding {<(1.4.2.4), (1.4.2.3)>}. Next, we locate the file streams D_3 and A_4. The codes of D_3 and A_4 are as follows:

D3 = {(1.2.1.2.2 0.6 0.8 0.5 0.4) (**1.2.2.0**.0 0.9 1 0.7 0.8) (**1.5.1.4**.2 0.8 0.7 1 0.9) …}
A3 = {(**1.2.2.0**.1 0.9 1 0.7 0.9) (**1.5.1.4**.1 0.8 0.7 1 0.7) (1.5.2.2.3 0.6 0.9 1 0.8) …}
L = 4

The query results that meet the requirements of structure and threshold are {<(1.2.2.0.0), (1.2.2.0.1)>, <(1.5.1.4.2), (1.5.1.4.1)>}.

We can see from the structure of the FST in Fig. 4.10, the query results above have the extension of negative twig, so we need to test the elements in the file streams of the negative leaf nodes. The code of the file stream E_1 is as follows:

E_1 = {(1.2.1.1 0.6 0.8 0.7) (**1.2.2.**1 1, 0.8 0.9) (0.6 0.9 0.8 1.5.2.3)…}
$L = 3$

The code of the element E_1 [0] is 1.2.1.1. There is no common ancestor in the first three layers in the Output, the pointer moves down, and finally we get the matching result is {<(1.5.1.4.2), (1.5.1.4.1)>}.

In order to deal with the fuzzy XML twig query with logic predicates, the methods of dealing with fuzzy twig query containing AND/OR/NOT-logic predicates are presented in this section, and also some analyses and query examples are provided. In the following section, we further discuss how to introduce the Index mechanism to improve the query efficiency.

4.3 Fuzzy XML Queries with Index

In order to improve the efficiency of querying XML, many researchers suggest to build Index on the algorithm. Many efforts on building index on classical XML can be found as introduced in Sect. 4.1. In this section, we introduce a fuzzy twig pattern query matching algorithm and further build index on the algorithm.

4.3.1 Fuzzy XML Twig Pattern Query Matching

Before we develop a fuzzy XML twig pattern query matching algorithm called FTwigStack, we first introduce some notions which will be used in the later algorithms.

Let Q be a twig pattern query and its root node is q. In general, q is used to indicate the twig pattern query whose root node is q also. The stack and data stream corresponding to the query node q in the twig pattern query are S_q and T_q, respectively. Moreover, regarding to the twig pattern query Q, the following node functions are defined:

1. *isLeaf*: Node Bool;
2. *isRoot*: Node Bool;
3. *Parent*: Node Node;

4. *children*: Node {Node};
5. *subtreeNodes*: Node {Node};
6. *getLastF*: Node Double;
7. *getSubPathF*: Node Double;
8. *getMergeF*: Node Double.

Here, *subtreeNodes* (*q*) returns node *q* and all its descendants. *getLastF* () returns the relative membership degree between the current node and its father node. *getSubPathF* () returns the whole membership degree of the path from the current node to the root of the query tree. *getMergeF* () returns the whole membership degree of the current query tree. Regarding to the stream *Tq*, the following functions are defined: *eof, advance, next, nextL* and *nextR*. Among them, *next* (*Tq*) gets the next node of the current node in the stream *Tq*; *advance* (*Tq*) makes the pointer of the stream *Tq* point to the next node; *nextL* (*Tq*) and *nextR* (*Tq*) functions respectively return *L* and *R* region coding value of the next node in *Tq*. Regarding to the stack *S*q, the following functions are defined: *empty, pop, push, topL* and *topR*. Here the last two functions respectively return the *L* and *R* region coding value of the top node in the stack. For an arbitrary query node *q* in query *Q*, its parent node in the query *Q* is *parent* (*q*) whose stack is *Sparent* (*q*).

Based on the existing fuzzy XML query algorithms in Liu et al. (2009) and Ma et al. (2011), the FTwigStack algorithm executes triple filtering in the process of getNext. It does not generate intermediate redundant nodes and each returned node forms the final solutions. When FTwigStack algorithm is used to query fuzzy twig with branches, the following three aspects should be considered:

1. determining whether the single path meets the requirements;
2. determining whether the whole path meets the requirements after the braches are combined, and the triple filtering will be done;
3. the output single path should be merged.

The algorithm FTwigStack is as follows.

The FTwigStack algorithm

```
1. While not end (q)
2.        q_act = getNext (q);
3.        if (! isRoot (q_act))
4.            cleanStack ( parent (q_act), nextL (q_act))
5.        if (isRoot (q_act) or !empty (S_parent(qact))
6.            cleanStack (q_act, nextL (q_act))
7.            moveStreamToStack (T_qact, S_qact, pointer to top (S_parent(qact)))
8.        if (isLeaf (q_act))
9.            showSolutionFromStack (S_qact, 0)
10.           pop (S_qact)
11.       else advance (T_qact)
12. mergeAllPathSolutions ()
```

The FTwigStack algorithm can be divided into two main stages. The first stage (lines 1–11) outputs all eligible nodes in twig pattern query in the order from the root to the leaves. Note that, although the output is a single path, each bifurcation node is processed through triple filtering. The first layer of filtering is to determine whether the membership degree between parent/child nodes meets the threshold requirement, the second layer of filtering is to determine whether the whole membership degree from the root node to the current node meets the threshold requirement, and the third layer of filtering is to determine whether the bifurcate nodes meet the threshold requirement after merging. All of these are handled through the getNex function. The second stage (line 12) merges the results obtained from the first stage and then gets the matching results satisfying the query conditions. Moreover, the getNext function is used so that each returned node has the extension of solution. Suppose the getNext (q) function returns a node q' in the query whose root is q. Then the following conditions must be satisfied:

1. it must have at least one extension of solution;
2. if q' has sibling nodes, then its sibling nodes must also have the extension of solution, and q' is the node whose left value is the smallest in all sibling nodes.

The following is the function getNext ().

The getNext () function

```
1.  if (isLeaf (q))
2.        while (getLastF (q) < w or getSubPathF (q) < w)
3.              advance (q)
4.        return q
5.  while (getLastF (q) < w or getSubPathF (q) < w)
6.        advance (q)
7.  for (q_i in children (q))
8.        n_i = getNext (q_i)
9.        If (n_i != q_i) return n_i
10. n_min = min arg_ni nextL (T_ni)
11. n_max = max arg_ni nextL (T_ni)
12. while (nextR (T_qi) < nextL (T_nmax) || getMergeF (T_q) < w)
13.       advance (T_q)
14. if (nextL (T_q) < nextL (T_nmin)) return q
15. else return n_min
```

In the function getNext (q), lines 1–4 first determine whether q is leaf node. If so, a eligible leaf node is searched and returned. Lines 5–6 conduct the first and second layers of filtering to the node q so that the first node which satisfies the first and second layer filter conditions can be found. Lines 7–9 recurrently call getNext () to return the eligible nodes in child nodes of the node q, in which it is ensured that every child has the extension of the solution. Lines 10–13 judge the file stream corresponding to the node q and find a node which satisfies the third filter conditions. In line 14, q is returned if q is the father of all child nodes returned.

Otherwise, a child node whose left value is smallest in child nodes is found as return (line 15).

We use an example to illustrate the FTwigStack algorithm. Figure 4.11 shows an XML document tree and query tree. The input of the FTwigStack algorithm is the file streams of the nodes *a, b, c* and *d*. The stacks S_a, S_b, S_c and S_d are created for these four nodes. In these stacks, the elements are saved in a 5-tuple: *left location, right location, depth, the result of calling the function subMerge* (), and *the pointer pointing to the parent node*. Here we assume that the threshold given by the user is 0.2. First, we call getNext () to determine whether there is a solution for *a*. Because *a* is not a leaf node, the values of the functions *getLastF* () and *getSubPathF* () of the node a_1 need to be judged. Since a_1 is the root node, both of the values of the two functions are 1 and the given conditions are satisfied. Next, we call getNext () for each child of *a*. Similarly, in getNext (*b*), we need to judge the values of the functions *getLastF* () and *getSubPathF* () of the current node b_1 in the file stream *b*. Both of the values are 0.9 and the first and second layers of filtering are met. Then it is needed to call getNex () for each child node of b_1. In getNext (*c*), since *c* is a leaf node, we need to find and return such a node that its values of the functions *getLastF* () and *getSubPathF* () meet conditions. Here the node c_1 meets the conditions. For getNext (*d*), since *d* is a leaf node, we also need to find a node in the file stream of *d*, which meets the conditions. Here d_1 meets the conditions. Finally, the node a_1 is pushed into the stack. The process of calling getNext in the second round is similar to the first round above, and b_1 is pushed into the stack. The third and fourth rounds return c_1 and d_1, respectively. Because c_1 and d_1 are leaf nodes, we output their path solutions [a_1, b_1, c_1] and [a_1, b_1, d_1]. The algoritm repeats until all the file streams move to the end. Finally, we get a solution (a_1, b_1, c_1, d_1) as shown in Fig. 4.12.

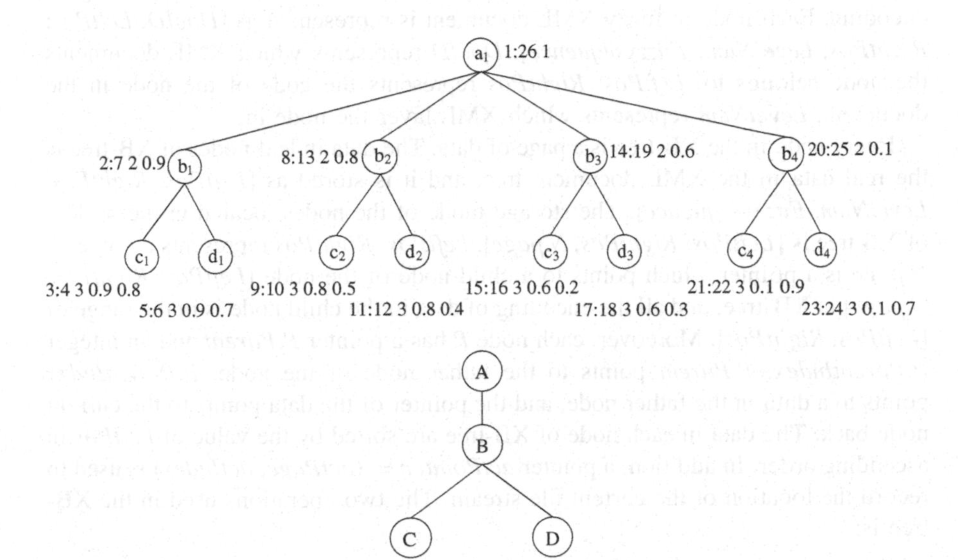

Fig. 4.11 A query example of the FTwigStack algorithm

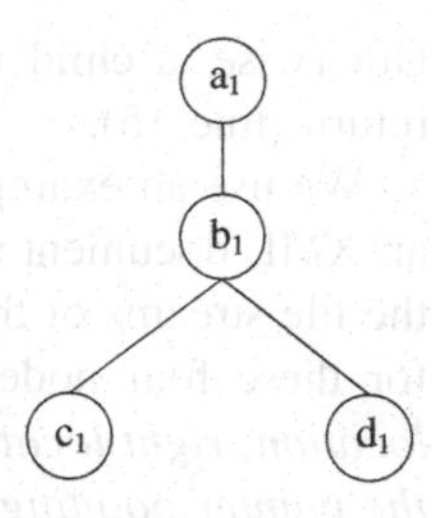

Fig. 4.12 The final query result of the example in Fig. 4.11

4.3.2 Fuzzy XML Twig Pattern Query with Index

The algorithm FTwigStack developed in Sect. 4.3.1 can deal with the fuzzy XML twig pattern query matching under the fuzzy XML environment. However, the FTwigStack algorithm must process all nodes in each file stream and does not make full use of the tree structure feature of XML. Assume that, for example, we know a node does not form the final solution in the structure. Then in the subtree whose root node is this given node, the nodes which have the same name as the given node do not form the final solution also. At this point, these nodes can be skipped directly and further process such as calculation and judgement are not needed. Therefore, we aim at further building the index on the FTwigStack algorithm in this section, and propose a fuzzy XML twig pattern query algorithm with index, called FTwigStackXB algorithm. According to the index mechnism, it can be determined that which nodes can be skipped directly and which parts need to be further processed, and thus the speed of query processing may be improved.

The FTwigStackXB algorithm is to build index on the FTwigStack algorithm. The building of index in fuzzy XML is based on the B-tree and the region encoding. Each node in fuzzy XML document is represented as [*DocID*, *LeftPos*: *RightPos*, *LevelNum*, *Fuzzysequence*]. *DocID* represents which XML documents the node belongs to. *LeftPos*: *RightPos* represents the code of the node in the document. *LevelNum* represents which XML layer the node in.

Every node in the XB-tree is a page of data. The data in leaf nodes of XB-tree is the real data in the XML document tree, and it is stored as [*LeftPos*: *RightPos*, *LevelNum*, *Fuzzysequence*]. The storage mode of the nodes located in the middle of XB tree is [*LeftPos*: *RightPos*, *N.page*]. *LeftPos*: *RightPos* represents a range *N*, *N.page* is a pointer which points to a child node of the node [*LeftPos*: *RightPos*, *N.page*] in XB-tree, and all the encoding of data in the child node is in the range of [*LeftPos*, *RightPos*]. Moreover, each node *P* has a pointer *P.Parent* and an integer *P.ParentIndex*. *P.Parent* points to the father node of the node. *P.ParentIndex* points to a data in the father node, and the pointer of the data points to the current node back. The data in each node of XB-tree are sorted by the value of *LeftPos* in ascending order. In addition, a pointer *actPointer* = (*actPage*, *actIndex*) is used to record the location of the current file stream. The two operations used in the XB-tree is:

1. *advance* () operation
   If the current pointer *actPointer* = (*actPage*, *actIndex*) does not point to the last data of the current node, the algorithm just moves *actIndex* forward. Otherwise it is needed to use (*actPage.parent*, *actPage.parentIndex*) instead of *actPointer* and do an *advance* operation.

2. *drilldown* () operation
   If *actPage* in the current pointer *actPointer* = (*actPage*, *actIndex*) is not a leaf node of XB-tree and *N* is the *actIndexth* node, it is needed to replace *actPointer* with (*N.page*, 0) and the current pointer points to the first child of *N.Page*.

Based on the FTwigStack algorithm, the FTwigStackXB algorithm builds index in the XML data stream. It can filter out more nodes that won't form the final solution, which further improves the efficiency of querying. When using the FTwigStackXB algorithm to process the twig pattern query, each node goes through triple filtering, and the node that the function getNext returns may not be a node in the final solution. It may be a data in XB-tree intermediate node. If so, the *advance* () or *drilldown* () operations may be proceed. The FTwigStackXB algorithm is given as follows.

The FTwigStackXB Algorithm

```
1. While not end (q)
2.      q_act = getNext (q);
3.      if (isPlainValue (T_qact))
4.           if (! isRoot (q_act))
5.                cleanStack (parent (q_act), nextL (q_act))
6.           if (isRoot (q_act) or !empty (S_parent(qact)))
7.                cleanStack (q_act, nextL (q_act) )
8.                moveStreamToStack (T_qact, S_qact, pointer to top
                  (S_parent(qact))
9.                if (isLeaf (q_act))
10.                    showSolutionFromStack (S_qact, 0)
11.                    pop (S_qact)
12.               else  advance (T_qact)
13.          else if (!isRoot (q_act) and empty (S_parent(qact)) or
                  nextL (T_parent(qact)) > nextR (T_qact))
14.               advance (T_qact)
15.          else drilldown (T_qact)
16. mergeAllPathSolutions ()
```

The algorithm FTwigStackXB mainly includes two stages: (1) lines 1–15 mainly work out the path solution conforming to the conditions. If the current node is the data of leaf nodes in XB-tree, then the current data node belongs to the nodes in final solution. Lines 4–8 do the operations of clearing and pushing the stack. Lines 9–12 output the path solution of the current node or move the pointer forward. Lines 13–14 mainly skip out the nodes that don't meet the conditions: if the current node isn't the root node and the parent node stack is empty, or the *L* of the current stream node of the father node is greater than the *R* of the current stream node, namely both of the current node and the nodes under the current node are unlikely to form final solutions, then skipping out them directly. Therefore, when building index, we should try to filter out more nodes through the line 14 of the main algorithm rather than turn into lines 4–12 and line 15 to judge. If the line 15 doesn't meet the conditions of the line 13, then move the current node down. (2) Line 16 mainly merges the path solutions derived from the first stage.

The getNext function in the algorithm FTwigStackXB is to return a node. If the node is a data of XB-tree intermediate nodes, i.e., it is not a node in final solution, then the lines 13–15 of the main algorithm are executed. If the node is a data of XB-tree leaf nodes, then the lines 4–12 of the main algorithm are executed. If the node is a data of leaf nodes, then the node has been triply filtered. The following is the function getNext ().

The getNext () Function

1. if (isLeaf (q))
2. while (getLastF (q) < w or getSubPathF (q) < w) advance (q)
3. return q
4. while (getLastF (q) < w or getSubPathF (q) < w)
5. advance (q)
6. for q_i in children (q)
7. n_i = getNext (q_i)
8. if (q != n_i or !isPlainValue (T_{ni})) return n_i
9. n_{min} = min arg$_{ni}$nextL (T_{ni})
10. n_{max} = max arg$_{ni}$nextL (T_{ni})
11. while (nextR (T_q) < nextL (T_{nmax}) || getMergeF (T_q) < w)
12. advance (T_q)
13. if (nextL (T_q) < nextL (T_{nmin})) return q
14. else return n_{min}

In the function getNext (), if the current node is a leaf node, then we filter out those nodes that don't satisfy the first and second layers of filter conditions, and return the first eligible leaf node (lines 1–3). If the current node isn't a leaf node, then we filter out those nodes that don't meet the first and second layers of filter conditions (lines 4–5). Next, each child of the current node recursively calls getNext (), if the returned result is not the current child node or the function is PlainValue () returns false, then we return the nodes that the file stream pointer points to (lines 6–8). Moreover, in lines 11–12, according to the scopes and the membership degrees of the child nodes of the current node, we filter out the nodes that don't meet the conditions. Also in line 13, if the current node is the father node of all child nodes, then the current node is returned, otherwise, the node whose left value is minimum in child nodes is returned.

In the following we provide an example to illustrate the differences between index and non-index in query. In brief, the FTwigStackXB algorithm can skip parts of data streams directly, but the FTwigStack algorithm needs to judge each data to determine whether it is one of the final solution. Figure 4.13 shows a fuzzy XML document and a query example.

For FTwigStack and FTwigStackXB algorithms, the input of the algorithms are the data stream file of each node in the query tree. The difference is when the FTwigStackXB algorithm executes the query, the algorithm builds B-tree index in the data stream file of each node. For the query tree and the fuzzy XML data file in Fig. 4.13, an XB-tree model can be created after building B-tree index in the data stream file of each node as shown in Fig. 4.14.

In Fig. 4.14, the four figures are the XB-index tree models building on the file streams of the nodes *A*, *B*, *C*, and *D*, respectively. At the beginning of the algorithm, the pointer of each file stream is as follows: *actA* = [12:25 pointer], *actB* = [3:10 pointer], *actC* = [4:21 pointer], and *actD* = [16:23 pointer]. When calling the getNext (*A*) function in round 1, the getNext (*B*) will be invoked, and the getNext (*B*) will continue to call getNext (*C*) and getNext (*D*). When calling the getNext (*C*), since *C* is a leaf node, *actC* = [4:21 pointer] will be returned. Moreover, because *actC* is not the data of the leaf nodes, *actC* will be returned to

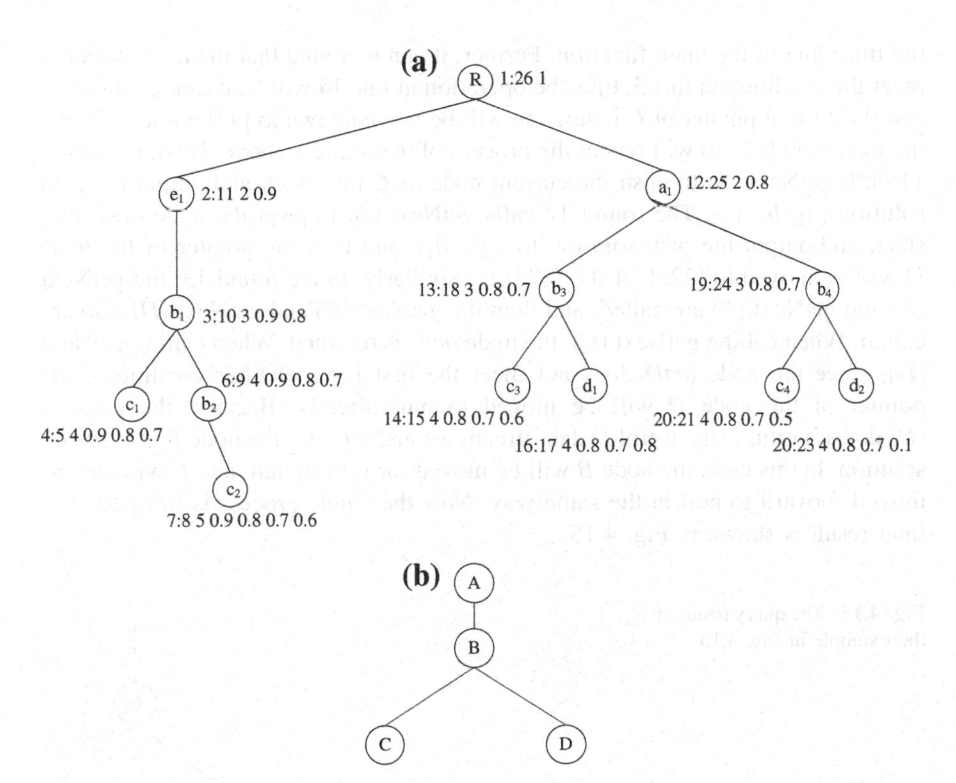

Fig. 4.13 A query example of the FTwigStackXB algorithm. **a** A fuzzy XML document with index. **b** A query tree

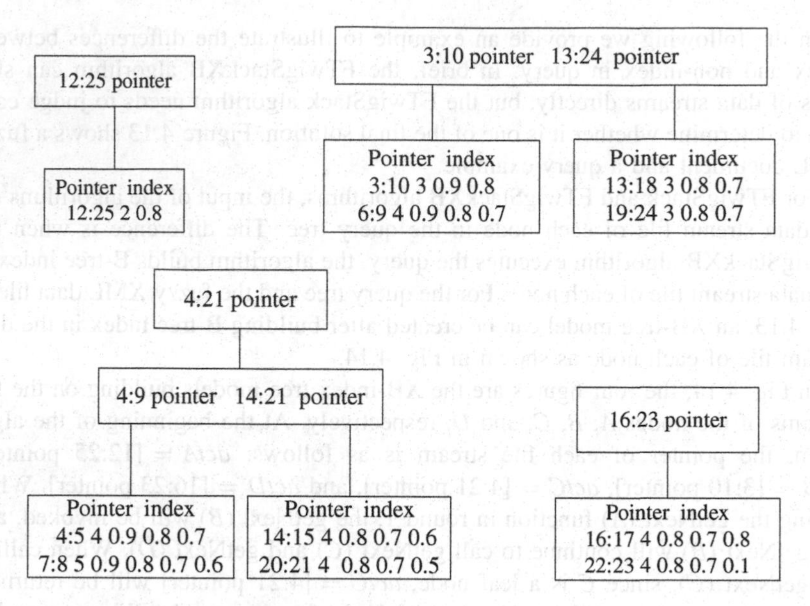

Fig. 4.14 An XB-tree model with B-tree index

the third line of the main function. Further, it can be found that the *actC* does not meet the condition in line 3, thus the operation in line 14 will be executed directly, and the current pointer of *C* index tree will be moved down to [4:9 pointer]. Then, the next rounds 2–10 will repeat the process of the round 1 above. Next, the round 11 calls getNext (*A*) to push the current node *actC* into stack and output the path solution [a_1, b_3, c_3]. The round 12 calls getNext (*A*) to push the node *actD* into stack and output the path solution [a_1, b_3, d_1], and then the pointer of the node *D* will be moved to [22:23 4 0.8 0.7 0.1]. Similarly, in the round 13, the getNext (*A*) and getNext (*B*) are called, and then the getNext (*C*) and getNext (*D*) also are called. When calling getNext (*C*), the node *actC* is returned. When calling getNext (*D*), since the node *actD* does not meet the first layer of filter conditions, the pointer of the node *D* will be moved to null directly. Because the getNext (*D*) doesn't return any useful *D* data stream for getNext (*B*), the node *B* also has no solution. In this case, the node *B* will be moved forward to null, and *C* will also be moved forward to null in the same way. Now the whole process is finished. The final result is shown in Fig. 4.15.

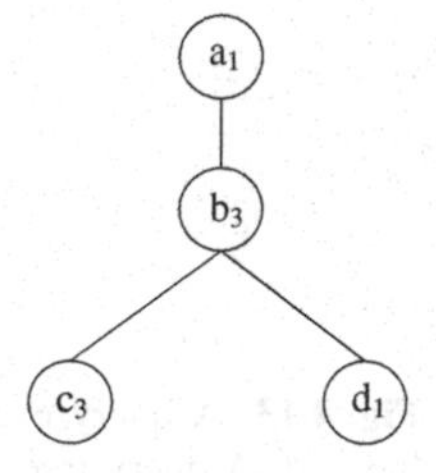

Fig. 4.15 The query result of the example in Fig. 4.13

4.4 Summary

The fuzzy XML data modelling has been extensively investigated as introduced in Chap. 3. Accordingly, how to query the fuzzy XML data is essential in order to meet the needs of practical applications. In particular, there are a lot of AND, OR and NOT logical predicates in the users' query conditions. So fuzzy XML query containing predicates becomes an important issue to be solved for fuzzy XML data management. Besides, how to ensure high efficiency is always the core problem of querying. Building index for fuzzy XML query can filter out the nodes which do not form final solutions and greatly improve the efficiency of querying. To this end, in this chapter, we present the methods of fuzzy XML twig queries with AND, OR and NOT predicates. Also we investigate the problem of fuzzy XML query based on index and built index on a fuzzy twig pattern query matching algorithm. The corresponding query examples are provided to well illustrate the proposed algorithms.

Besides the querying of fuzzy XML data, with the popularity of Web-based applications, the requirement has been put on the exchange and share of data among various applications. For example, currently there is an increasing need to effectively publish fuzzy structured data in fuzzy database models (such as fuzzy relational database models and fuzzy object-oriented database models) as fuzzy XML documents for Web-based applications or reengineer fuzzy XML data into fuzzy database models to satisfy the needs of storing fuzzy XML data in fuzzy databases. All of these will be introduced in the later chapters.

References

Abiteboul S, Senellart P (2006) Querying and updating probabilistic information in XML. In: Proceedings of EDBT, pp 1059–1068

André Y, Caron AC, Debarbieux D, Roos Y (2005) Indexes and path constraints in semistructured data. In: Proceedings of the sixteenth international workshop on database and expert systems applications, pp 837–841

Bruno N, Koudas N, Srivastava D (2002) Holistic twig joins: optimal XML pattern matching. In: Proceedings of SIGMOD, pp 310–321

Chung CW, Min JK, Shim K (2002) APEX: an adaptive path index for XML data. In: Proceedings of the 2002 ACM SIGMOD international conference on management of data, pp 121–132

Garakani V, Izadi SK, Haghjoo M, Harizi M (2007) Ntjfsat¬: a novel method for query with not-predicates on xml data. In: Proceedings of the sixteenth ACM conference on conference on information and knowledge management, pp 885–888

Hung E, Getoor L, Subrahmanian VS (2003) PXML: a probabilistic semistructured data model and algebra. In: Proceedings of ICDE, pp 467–478

Kimelfeld B, Sagiv Y (2007) Matching twigs in probabilistic XML. In: Proceedings of VLDB, pp 27–38

Kimelfeld B, Kosharovshy Y, Sagiv Y (2008) Query efficiency in probabilistic XML models. In: Proceedings of SIGMOD, pp 701–714

Li Q, Moon B (2001) Indexing and querying XML data for regular path expressions. In: VLDB, pp 361–370

Li Y, Wang G, Xin J, Zhang E, Qiu Z (2009) Holistically twig matching in probabilistic XML. In: Proceedings of ICDE, pp 1649–1656

Liu J, Ma ZM, Yan L (2009) Efficient processing of twig pattern matching in fuzzy XML. In: Proceedings of the 18th ACM conference on information and knowledge management, pp 193–204

Lu J, Ling TW, Chan CY, Chen T (2005a) From region encoding to extended dewey: on efficient processing of XML twig pattern matching. In: Proceedings of VLDB, pp 193–204

Lu J, Ling TW, Chan CY, Chen T (2005b) From region encoding to extended dewey: on efficient processing of xml twig pattern matching. Technical report, TRA6/05 National university of Singapore, Singapore

Lu J, Meng X, Ling TW (2011) Indexing and querying XML using extended Dewey labeling scheme. Data Knowl Eng 70(1):35–59

Ma ZM, Liu J, Yan L (2011) Matching twigs in fuzzy XML. Inf Sci 181(1):184–200

Rao P, Moon B (2004) PRIX: indexing and querying XML using prufer sequences. In: Proceedings of the 20th international conference on data engineering, pp 288–299

Senellart P, Abiteboul S (2007) On the complexity of managing probabilistic XML data. In: Proceedings of PODS, pp 283–292

Tatarinov I, Viglas S, Beyer KS, Shanmugasundaram J, Shekita EJ, Zhang C (2002) Storing and querying ordered XML using a relational database system. In: Proceedings of SIGMOD, pp 204–215

Wang H, Park S, Fan W, Yu PS (2003) ViST: a dynamic index method for querying XML data by tree structures. In: Proceedings of the 2003 ACM SIGMOD international conference on management of data, pp 110–121

Zadeh LA (1965) Fuzzy sets. Inf Control 8(3):338–353

Chapter 5
Fuzzy XML Extraction from Fuzzy Database Models

Abstract Nowadays most of data are modeled by database modes such as UML, relational and object-oriented database models as introduced in Chap. 1. Then database administrators are faced with the challenge of ensuring their databases to interface with other heterogeneous systems using XML, which is the de facto standard for publishing and exchanging data on the Web. However, information is often imprecise and uncertain in many real-world applications, and thus fuzzy database models and fuzzy XML have been extensively investigated as introduced in Chaps. 2 and 3. Accordingly, there is an increasing need to automate the process of extracting fuzzy XML models containing information from existing fuzzy database models. In this chapter, we introduce how to extract fuzzy XML from several typical fuzzy database models, including fuzzy UML data models, fuzzy relational database models, and fuzzy object-oriented database models.

5.1 Introduction

As introduced in Chap. 1, databases and XML are the important techniques for manage data in real-world applications. In particular, XML has become the de facto standard for publishing and exchanging data on the Web. Currently, most of the data are stored in databases such as relational and object-oriented databases. Thus in order to realize the full potential of XML, it is of significance to extract the XML documents from existing databases (Fernández et al. 2000). Liu et al. (2006) provided a constraint preserving transformation from relational schema to XML schema. Fong and Cheung (2005) proposed an approach for transforming a relational schema into an XML Schema definition for creating an XML database. Lo et al. (2010) presented a method to query and integrate relational databases to produce XML documents and the corresponding schemas. Naser et al. (2009) presented a novel approach for mapping an existing object-oriented database into XML and vice versa. Wang et al. (2005) developed a system named COCALEREX (converting relational to XML) to handle the transformation process for both catalog-based and legacy relational databases.

L. Yan et al., *Fuzzy XML Data Management*, Studies in Fuzziness and Soft Computing 311, DOI: 10.1007/978-3-642-44899-7_5,

In many real-world applications, information is often imprecise and uncertain, and thus fuzzy data modeling has been extensively investigated in various database models, which has resulted in numerous contributions, mainly with respect to the popular fuzzy conceptual data models (fuzzy ER/EER model, fuzzy UML model, and etc.) and fuzzy logical database models (fuzzy relational database model, fuzzy object-oriented database model, and etc.) as introduced in Chap. 2. Also, topics related to the modeling of fuzzy data can be considered very interesting in the XML data context, and fuzzy XML models have been developed as introduced in Chap. 3. As such, there is an increasing need to effectively publish fuzzy structured data in fuzzy database models as fuzzy XML documents for Web-based applications. Ma and Yan (2007) investigated the formal conversions from the fuzzy UML model to the fuzzy XML model and developed the fuzzy UML data model to design the fuzzy XML model conceptually. Yan et al. (2011) proposed a formal conversion approach from the fuzzy relational databases to the fuzzy XML model, which is established by introducing a series of mapping rules, and demonstrated the validity of the proposed approach by using a practical case. The conversions from the fuzzy relational databases to the fuzzy XML model lay a theoretical foundation for establishing the overall management system of fuzzy XML data. Liu and Ma (2013) studied how to automatically generate fuzzy XML documents containing information from existing fuzzy object-oriented databases. In this chapter, we introduce how to extract fuzzy XML from fuzzy database models including fuzzy UML data models, fuzzy relational database models, and fuzzy object-oriented database models.

5.2 Fuzzy XML Extraction from Fuzzy UML Data Models

In order to extract fuzzy XML from fuzzy UML data models, Ma and Yan (2007) investigated the formal conversions from the fuzzy UML data model to the fuzzy XML model and developed the fuzzy UML data model to design the fuzzy XML model conceptually. For the transformation approach, relevant constructs are the fuzzy extensions of those of UML's Static View, consisting of the fuzzy classes and their relationship such as fuzzy association, fuzzy generalization, and various kinds of fuzzy dependencies as introduced in Sect. 2.3. We develop the transformation of these constructs into DTD fragments.

5.2.1 Extraction of Fuzzy Classes

UML classes are transformed into XML element type declarations (Conrad et al. 2000). Here, the class names become the names of the element types and the attributes are transformed into element content description. It is noted that, in the UML, attribute names are mandatory, whereas the attribute types are optional. In contrast, an element content only consists of type names in the XML. As a result, it is assumed that attribute names imply their attribute type names (Conrad et al. 2000). When there is no class representing a suitable declaration for an attribute type, the attribute type is assumed to be an element whose content type is #PCDATA. In addition, multiplicity specifications of attributes are mapped into cardinality specifications with specifiers ?, *, and +, which are used for element content construction.

In the fuzzy UML model, four kinds of classes can be identified, which are

(a) classes without any fuzziness at the three levels,
(b) classes with fuzziness only at the third level,
(c) classes with fuzziness at the second level, and
(d) classes with fuzziness at the first level.

For the classes in case (a), they can be transformed following the approach developed in Conrad et al. (2000). The transformation of the classes with the third and second levels of fuzziness is of particular concern. Instead of formal definitions, in the following we utilize some examples to illustrate how to transform the classes with the third and second levels of fuzziness into XML DTD.

First, let us look at class "student" in Fig. 5.1.

It is clear that this class has two attributes, "age" and "e-mail," taking fuzzy values representing various possible distributions. In other words, the class has the third level of fuzziness. While the class name becomes the name of the element type, and the attributes are transformed into element content description, these two attributes cannot be directly transformed into the element content description with content type #PCDATA. We should use

```
<!ELEMENT age (Dist)>
<!ELEMENT Dist (Val+)>
  <!ATTLIST Dist type (disjunctive)>
<!ELEMENT Val (#PCDATA)>
  <!ATTLIST Val Poss CDATA "1.0">
```

rather than use

```
<!ELEMENT age (#PCDATA)>.
```

Similarly, we use

```
<!ELEMENT email (Dist)>
<!ELEMENT Dist (Val+)>
  <!ATTLIST Dist type (conjunctive)>
<!ELEMENT Val (#PCDATA)>
  <!ATTLIST Val Poss CDATA "1.0">
```

in place of

```
<!ELEMENT e-mail (#PCDATA)>.
```

student

sname
FUZZY age
Sex
FUZZY e-mail

```
<!ELEMENT student (sname?, age?, sex?, email?)>
        <!ATTLIST student SID IDREF #REQUIRED>
<!ELEMENT sname (#PCDATA)>
<!ELEMENT age (Dist)>
<!ELEMENT Dist (Val+)>
        <!ATTLIST Dist type (disjunctive)>
<!ELEMENT sex (#PCDATA)>
<!ELEMENT email (Dist)>
<!ELEMENT Dist (Val+)>
        <!ATTLIST Dist type (conjunctive)>
<!ELEMENT Val (#PCDATA)>
        <!ATTLIST Val Poss CDATA "1.0">
```

employee

fname
position
office
course
μ

```
<!ELEMENT employee (Dist)>
        <!ATTLIST employee FID IDREF #REQUIRED>
<!ELEMENT Dist (Val+)>
        <!ATTLIST Dist type (disjunctive)>
<!ELEMENT Val (fname?, position?, office?, course?)>
        <!ATTLIST Val Poss CDATA "1.0">
<!ELEMENT fname (#PCDATA)>
<!ELEMENT position (#PCDATA)>
<!ELEMENT office (#PCDATA)>
<!ELEMENT course (#PCDATA)>
```

Fig. 5.1 Transformation of the classes in the fuzzy UML to the fuzzy XML

Now let us focus on class "employee" in Fig. 5.1. This class has the second level of fuzziness. That means that the class instances belong to the class with membership degrees. For such a class, when its class name becomes the name of the element type, the attributes cannot be transformed into element content description directly. We should use

```
<!ELEMENT employee (Dist)>
    <!ATTLIST employee FID IDREF #REQUIRED>
<!ELEMENT Dist (Val+)>
    <!ATTLIST Dist type (disjunctive)>
<!ELEMENT Val (fname?, position?, office?, course?)>
    <!ATTLIST Val Poss CDATA "1.0">
```

rather than directly use

```
<!ELEMENT employee (fname?, position?, office?, course?)>
    <!ATTLIST employee FID IDREF #REQUIRED>.
```

Figure 5.1 depicts the details transforming classes "student" and "employee" into the fuzzy XML. An aggregation represents a whole-part relationship between an aggregate and a constituent part. We can treat all constituent parts as the special attributes of the aggregate. Then we can transform the aggregations using the approach to the transformation of classes.

5.2.2 *Extraction of Fuzzy Generalizations*

The generalization in the UML defines a subclass/superclass relationship between classes: one class, called superclass, is a more general description of a set of other classes, called subclasses. Following the same transformation of classes given above, the superclass and each subclass are all transformed into the element types in the XML, respectively. Here the element type originating from the superclass is called a *superelement* and the element type originating from a subclass is called a *subelement* in Conrad et al. (2000). Note that a superelement must receive an additional ID attribute stated #REQUIRED, and each subelement must be augmented by a #REQUIRED IDREF attribute in addition to the transformations that the class names become the names of the element types and the attributes are transformed into element content description.

Now consider the fuzziness in the generalization in the fuzzy UML model. Assume that the superclass and subclasses involved in the generalization may have fuzziness at the type/instance level (the second level) and/or at the attribute value

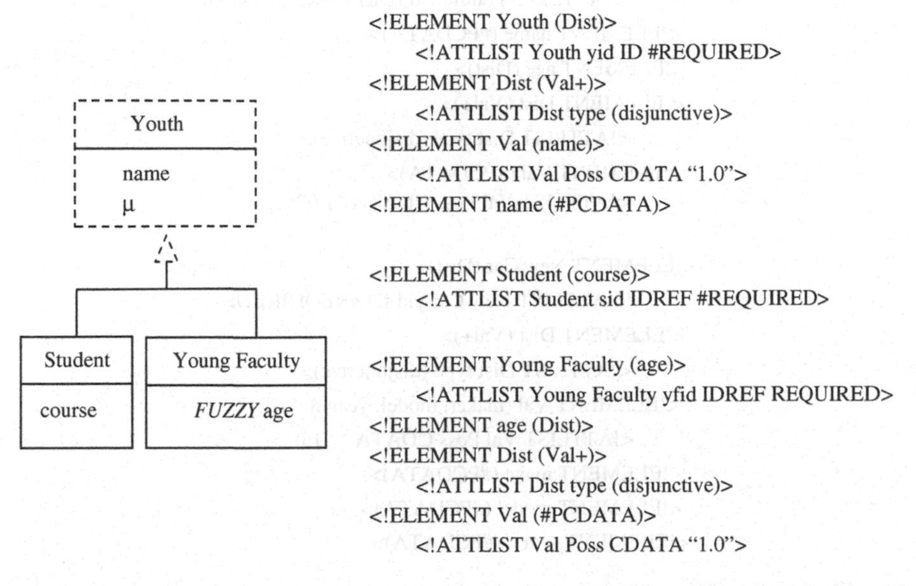
Youth
name
μ
Student
course
Young Faculty
FUZZY age
<!ELEMENT Youth (Dist)>
<!ATTLIST Youth yid ID #REQUIRED>
<!ELEMENT Dist (Val+)>
<!ATTLIST Dist type (disjunctive)>
<!ELEMENT Val (name)>
<!ATTLIST Val Poss CDATA "1.0">
<!ELEMENT name (#PCDATA)>
<!ELEMENT Student (course)>
<!ATTLIST Student sid IDREF #REQUIRED>
<!ELEMENT Young Faculty (age)>
<!ATTLIST Young Faculty yfid IDREF REQUIRED>
<!ELEMENT age (Dist)>
<!ELEMENT Dist (Val+)>
<!ATTLIST Dist type (disjunctive)>
<!ELEMENT Val (#PCDATA)>
<!ATTLIST Val Poss CDATA "1.0">

Fig. 5.2 Transformation of the generalizations in the fuzzy UML to the fuzzy XML

level (the third level). The transformation of such superclass and subclasses can be finished according to the transformation of fuzzy classes developed above. Meanwhile, the created superelement and each subelement must be associated with ID #REQUIRED and IDREF #REQUIRED, respectively. Figure 5.2 depicts the transformation of the fuzzy generalization.

5.2.3 Extraction of Fuzzy Associations

Associations are relationships that describe connections among class instances. An association is a more general relationship than aggregation or generalization. So basically we can transform the associations in the UML using the approach to the transformation of generalizations given above. That is, first the class names

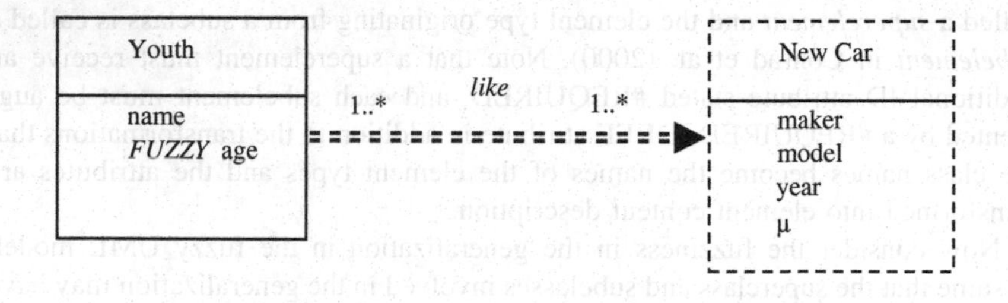


```
<!ELEMENT Youth (name, age)>
        <!ATTLIST Youth ncid IDREF #REQUIRED>
<!ELEMENT name (#PCDATA)>
<!ELEMENT age (Dist)>
<!ELEMENT Dist (Val+)>
        <!ATTLIST Dist type (disjunctive)>
<!ELEMENT Val (#PCDATA)>
        <!ATTLIST Val Poss CDATA "1.0">

<!ELEMENT New Car (Dist)>
        <!ATTLIST New Car yid ID #REQUIRED>
<!ELEMENT Dist (Val+)>
        <!ATTLIST Dist type (disjunctive)>
<!ELEMENT Val (maker, model, year)>
        <!ATTLIST Val Poss CDATA    "1.0">
<!ELEMENT make (#PCDATA)>
<!ELEMENT model (#PCDATA)>
<!ELEMENT year   (#PCDATA)>
```

Fig. 5.3 Transformation of the associations in the fuzzy UML to the fuzzy XML

become the names of the element types and the attributes are transformed into element content description. Then each element transformed must be augmented by a #REQUIRED IDREF attribute (ISIS XML/EDI Project 2001), which is an artificial one and from another class involved in the association.

Since, in the fuzzy UML data model, each class involved in an association may have fuzziness at the type/instance level (the second level) and/or at the attribute value level (the third level), its transformation must be carried out according to the transformation of fuzzy classes developed above. Utilizing this approach, Fig. 5.3 depicts the transformation of the fuzzy association.

5.3 Fuzzy XML Extraction from Fuzzy Relational Database Models

For managing fuzzy XML data, Yan et al. (2011) investigated the formal conversions from the fuzzy relational databases to the fuzzy XML model, which is established by introducing a series of mapping rules. By using a practical case, we finally demonstrated the validity of the proposed approach. The conversions from the fuzzy relational databases to the fuzzy XML model lay a theoretical foundation for establishing the overall management system of fuzzy XML data.

In order to realize the transformation of relational database into XML, Lee et al. (2001) defined the nest operator, which can transform the relational database into the non-1NF nesting relational database. The nest operator is defined: Let r be a n-ary table with column set $C, A \in C$ and $\bar{A} \in C - A$. For each $(n - 1)$-tuple $r \in \prod_{\bar{A}}(r)$, defining an n-tuple t' as follows:

$$\left.\begin{array}{c} t'[\overline{\mathrm{A}}] = t \\ t'[\mathrm{A}] = \{s[A] \| s \in r \wedge s[\overline{A}] = t\} \end{array}\right\} then,\ nest_{\mathrm{A}}(r) = \{t' | t \in \Pi_{\bar{\mathrm{A}}}(r)\}$$

After $nest_{\mathrm{A}}(r)$, if the attibute A has only a set with single value $\{v\}$, then we say that nesting failed and treat $\{v\}$ and v interchangeably, i.e., $\{v\} = v$. Therefore, when nesting failed, the following is true: $nest_{\mathrm{A}}(r) = r$. Otherwise, if the attibute A has a set with multiple values $\{v_1, \ldots, v_k\}$ $(k \geq 2)$, we say that nesting sucessful.

The conversions from the fuzzy relational databases to the fuzzy XML model are completed through two steps: the first is to transform the relational database into the XML DTD regardless of the existence of fuzzy information, and then modify the DTD based on the fuzzy information in the relational tables. The conversions are established by a series of mapping rules as follows.

Rule 1: for a relational database schema, a root node is created in XML and the corresponding DTD is described as:

<!ELEMENT root (element*)>.

Rule 2: for each relation, creating a non-leaf element node in DTD.
Rule 3: for a primary key, it is declared as:

<!ATTLIST Ename Aname ID…>.

Rule 4: for a foreign key, if the foreign key refers to the single ID, it is declared as:

<!ATTLIST Ename Aname IDREF…>;

Otherwise, if the foreign key refers to multiple IDs, it is declared as:

<!ATTLIST Ename Aname IDREFS…>.

Rule 5: for the attribute columns which are not the primary keys and foreign keys, they are declared as:

<!ELEMENT Ename (original-definition)>

Rule 6: for the constraint "NOT NULL" of the attribute value, it is represented by:

<!ALLIST Ename Aname original-definition #REQUIRED>

Rule 7: for a relation whose primary key does not include foreign keys or whose primary key includes at least two foreign keys, a DTD declaration is created below the root element:

<!ELEMENT root (element*)>

Rule 8: for a relation r_1 whose primary key includes a foreign key and its parent relation is r_2, the relation r_1 is transformed into the sub-relation of r_2:

<!ELEMENT element_2 (element_1*)>

Rule 9: for the relations r_1 and r_2, if there is only a $N - 1$ relation between r_1 and r_2, the relation r_1 is transformed into the sub-relation of r_2:

<!ELEMENT element_2 (element_1*)>

Rule 10: if there are multiple $N - 1$ relations from the relation r_0 to the relations $r_1,\ldots, r_k$, the relation r_0 is transformed into the sub-relation of $r_1,\ldots, r_k$, respectively:

<!ELEMENT element_1 (element_0*)>
…
<!ELEMENT element_k (element_0*)>

Rule 11: If there is a $N - N$ relation between two relations r_1 and r_2, the relations r_1 and r_2 are transformed into the sub-element of the root element:

<!ELEMENT root (element_1*, element_2*)>

And then we use the attribute declarations ID and IDREF as introduced above to describe element_1 and element_2.

Rule 12: if there is a $N - N$ relation in r_1, the relation r_1 is transformed into the sub-element of the root element:

<!ELEMENT root (element$_1$*)>

And then we use the attribute declaration ID to describe element$_1$.

Rule 13: for a relation r, it is represented as r (A_1,…, A_{k-1}, A_k,…, A_n) after the nest operator, where the attribute (A_1,…, A_{k-1}) is a nesting structure. If $k = 1$, i.e., there are not nesting relations, the attribute columns are transformed into sub-elements:

<!ELEMENT Ename (original-definition)>;

If $k > 1$, i.e., there are nesting relations, then:

- for each attribute A_i ($1 \le i \le k - 1$), if A_i may be null, the element content can be represented as A_i* or A_i+, i.e.,

  <!ELEMENT Ename (Aname$_1$*, Aname$_2$+,…)>

- for each attribute A_j ($k \le j \le n$), if A_j may be null, the element content can be represented as A_j? or A_j, i.e.,

  <!ELEMENT Ename (…, Aname$_1$?, Aname$_2$)>

Rule 14: for the constraint "default" of the attribute, it is represented by:

<!ALLIST element Aname original-definition "default">

By using the 14 rules above, the relational databases can be transformed into XML DTD. The transformation algorithm is briefly given as follows:

(a) Regarding to the attribute columns not containing fuzzy information, the nest operator is defined on the other attribute columns in the relation;
(b) By applying Rule 1, the root element of the DTD tree is created;
(c) By applying the Rules 2 and 7, a suitable relation called r_1 is found to create the sub-elements of the root node;
(d) For the attribute columns of the relation r_1, the corresponding DTD declarations are created with the Rules 3, 4, 5, 6, 13 and 14;
(e) For the relations r_2,…, r_n which refer to the relation r_1, the corresponding sub-elements of r_1 are created with the Rules 8, 9, 10, 11, and 12;
(f) The rest relations are traversed recursively and the corresponding parent elements are found with the Rules 8, 9, 10, 11, and 12.

Next we consider the fuzzy information in the relations. At this point it may be needed to modify the created DTD above. When the relation r contains the fuzzy information, the following rules are used.

Rule 15: for the relation containing the attribute $\mu \in (0, 1]$, which represents the membership degree of a tuple belonging to the relation, it is transformed as follows:

- creating a non-leaf node within the DTD sub-element corresponding to the relation *r*, i.e., *Val* sub-element node, and the number of occurrences of the element is defined as +:

  <!ELEMENT element (Val+)>

- creating the content of the attribute columns of the relation *r* within the *Val* sub-element, which can be similarly processed with the transformations without fuzzy value, and the default value of *Val* is 1.0:

  <!ELEMENT Val (element$_1$*, element$_k$+, element$_1$?, element$_1$,…)>
  <!ATTLIST Val Poss CDATA "1.0">

Rule 16: for the relation whose attribute values are represented by possibility distributions, if the created DTD is the leaf node sub-elements, it is transformed as follows:

- creating a sub-element node *Dist* within the DTD sub-element corresponding to the relation *r*:

  <!ELEMENT element (Dist)>

- creating the sub-element *Val* within the sub-element node *Dist*, and the number of occurrences of the element is defined as +:

  <!ELEMENT Dist (Val+)>

- creating the content of the attribute columns of the relation *r* within the *Val* sub-element, which can be similarly processed with the transformations without fuzzy value, and the default value of *Val* is 1.0:

  <!ELEMENT Val (original-definition)>
  <!ATTLIST Val Poss CDATA "1.0">

Rule 17: for the relation whose attribute values are represented by possibility distributions, if the created DTD is the non-leaf node sub-elements, it is transformed as follows:

- creating a sub-element node *Dist* within the DTD sub-element corresponding to the relation *r*:

  <!ELEMENT element (Dist)>

- creating the sub-element *Val* within the sub-element node *Dist*, and the number of occurrences of the element is defined as +:

  <!ELEMENT Dist (Val+)>

- creating the content of the attribute columns of the relation *r* within the *Val* sub-element, which can be similarly processed with the transformations without fuzzy value, and the default value of *Val* is 1.0:

<!ELEMENT Val (element$_1$*, element$_k$+, element$_l$?, element$_l$, …)>
<!ATTLIST Val Poss CDATA "1.0">

In the following, we provide a transformation example from fuzzy relational database to fuzzy XML DTD based on the Rules 1–17. Given the relations of *University*, *Employee*, *Department* and *Student* as shown in the Tables 5.1, 5.2, 5.3 and 5.4, their primary keys are *Uname*, *Dname*, *EID* and SID, respectively. Also *Uname* is the foreign key of the relation *Department*, and *Dname* is the foreign key of the relations *Employee* and *Student*. In these tables, the primary keys of the relations are shown in boldface and the foreign keys are italics.

Firstly, we can create a simple DTD without fuzzy values based on the rules 1–14. After creating the root node element, we find a suitable table in the relation tables as the sub-element of the root. Here *University* does not have the foreign keys and it can be the sub-element of the root node. Then we create the DTD directly below the root node. Here we take no account of the fuzzy attribute column *PD*. Next, we find the table *Department* which refers to the primary key *Uname* in *University*, and create the corresponding sub-elements below the element *University*. Since the tables *Employee* and *Student* refer to the attribute column *Dname* in *Department*, they are transformed as the sub-elements of *Department*. The created DTD is as follows:

```
<!ELEMENT root (university*)>
<!ELEMENT university (address+, department*)>
  <!ATTLIST university Uname ID #REQUIRED>
<!ELEMENT department (location+, employee*, student*)>
  <!ATTLIST department Dname ID #REQUIRED>
  <!ATTLIST department Uname IDREF #REQUIRED>
<!ELEMENT employee (ename?, position?, office?)>
  <!ATTLIST employee EID ID #REQUIRED>
  <!ATTLIST employee Dname IDREF #REQUIRED>
<!ELEMENT student (sname?, sex?, age?)>
  <!ATTLIST student SID ID #REQUIRED>
  <!ATTLIST student Dname IDREF #REQUIRED>
<!ELEMENT address (#PCDATA)>
<!ELEMENT location (#PCDATA)>
<!ELEMENT ename (#PCDATA)>
<!ELEMENT position (#PCDATA)>
<!ELEMENT office (#PCDATA)>
<!ELEMENT sname (#PCDATA)>
<!ELEMENT sex (#PCDATA)>
<!ELEMENT age (#PCDATA)>
```

Then, we further modify the created DTD by considering the fuzzy information based on the Rules 15–17:

Table 5.1 A fuzzy relation *university*

Uname	Address	PD
Oakland university	Detroit	0.8
Wayne state university	Detroit	1.0

Table 5.2 A fuzzy relation *employee*

EID	Dname	Ename	Position	Office	PD
85431095	Computer science and engineering	Frank Yager	Associate professor	B1024	0.8
	Computer science and engineering	Frank Yager	Professor	B1024	0.6

Table 5.3 A fuzzy relation *department*

Dname	Uname	Location
Computer science and engineering	Oakland university	Oakland county

Table 5.4 A fuzzy relation *student*

SID	Dname	Sname	Sex	Age
20023056	Computer science and engineering	Tom Smith	Male	Young

1. The fuzzy attribute *PD* in table *University* is a single membership degree. According to Rule 15, we have:

```
<!ELEMENT university (address+, Val+)>
<!ELEMENT Val (department*)>
  <!ATTLIST Val Poss CDATA “1.0”>
```

2. The attribute age in *Student* is a possibility distribution, and it is a leaf node sub-element. According to Rule 16, we have:

```
<!ELEMENT age (Dist)>
<!ELEMENT Dist (Val+)>
<!ELEMENT Val (#PCDATA)>
  <!ATTLIST Val Poss CDATA “1.0”>
```

3. The fuzzy attribute PD in the table Employee is a multiple membership degrees, and it is a non-leaf node sub-element. According to Rule 17, we have:

```
<!ELEMENT element (Dist)>
<!ELEMENT Dist (Val+)>
<!ELEMENT Val (ename?, position?, office?)>
  <!ATTLIST Val Poss CDATA “1.0”>
```

After modifying, the final DTD with fuzzy information is as follows:

```
<!ELEMENT root (university*)>
<!ELEMENT university (address+, Val*)>
```

```
    <!ATTLIST university Uname ID #REQUIRED>
  <!ELEMENT Val (department*)>
    <!ATTLIST Val Poss CDATA "1.0">
  <!ELEMENT department (location+, employee*, student*)>
    <!ATTLIST department Dname ID #REQUIRED>
    <!ATTLIST department Uname IDREF #REQUIRED>
  <!ELEMENT employee (Dist)>
    <!ATTLIST employee EID ID #REQUIRED>
    <!ATTLIST employee Dname IDREF #REQUIRED>
  <!ELEMENT Dist (Val+)>
  <!ELEMENT Val (ename?, position?, office?)>
    <!ATTLIST Val Poss CDATA "1.0">
  <!ELEMENT student (sname?, sex?, age?)>
    <!ATTLIST student SID ID #REQUIRED>
    <!ATTLIST student Dname IDREF #REQUIRED>
  <!ELEMENT address (#PCDATA)>
  <!ELEMENT location (#PCDATA)>
  <!ELEMENT ename (#PCDATA)>
  <!ELEMENT position (#PCDATA)>
  <!ELEMENT office (#PCDATA)>
  <!ELEMENT sname (#PCDATA)>
  <!ELEMENT sex (#PCDATA)>
  <!ELEMENT age (Dist)>
  <!ELEMENT Dist (Val+)>
  <!ELEMENT Val (#PCDATA)>
  <!ATTLIST Val Poss CDATA "1.0">
```

5.4 Fuzzy XML Extraction from Fuzzy Object-Oriented Database Models

There are some similarities between the notions in fuzzy object-oriented database models and fuzzy UML models, e.g., the notions fuzzy classes, fuzzy relationships and fuzzy inheritance hierarchies in fuzzy object-oriented database models are similar with the notions fuzzy classes, fuzzy associations and fuzzy generalizations in fuzzy UML models as mentioned in Chap. 2. Therefore, the approach for extracting fuzzy XML from fuzzy object-oriented database models may be given following the similar procedure of extracting fuzzy XML from fuzzy UML models (Ma and Yan 2007) as introduced in Sect. 5.2. In the following, to publish fuzzy data in fuzzy object-oriented database models as fuzzy XML documents for Web-based applications, we introduce how to extract fuzzy XML from fuzzy object-oriented database models.

As introduced in Sect. 2.5, in general, a fuzzy object-oriented database (FOODB) model consists of the basic notions including fuzzy object, fuzzy class, fuzzy relationship, and fuzzy inheritance. In a FOODB model, an object may belong to a class with a membership degree of [0, 1] and a class may be the subclass of another class with membership degree of [0, 1]. Also, a class consists of a class name, attributes (including a unique object identity *OID* attribute), and methods. The domains of some attributes may be fuzzy, and in this case the attributes are called fuzzy attributes, where a fuzzy keyword *FUZZY* is appeared in front of an attribute indicating the attribute may take fuzzy values. In the following, we introduce how to transform the main notions of a FOODB model into fuzzy XML. Here, the transformed fuzzy XML is represented based on the fuzzy XML DTD, and the fuzzy XML Schema can be extracted similarly.

5.4.1 Extraction of Fuzzy Classes

A class consists of a class name, attributes, and methods. As mentioned in Conrad et al. (2000), UML classes are transformed into XML element type declarations. Based on the idea, a class in a FOODB model is also transformed into a fuzzy XML element type declaration. In the following, we start by introducing some transformation rules for extracting fuzzy XML from fuzzy classes in FOODB models.

Rule 1: For each class *FC* in a FOODB model, an element type *FC* is created, which has the same name with the class name in the FOODB model.

Rule 2: For each attribute *FA* of a fuzzy class *FC*, if the attribute is an object identity attribute, it is transformed into an attribute description of the element type *FC*; otherwise, it is transformed into the element content description of the element *FC*. Here, the name of an attribute provides the name for the element type in content description.

Rule 3: For each class $FC = [ID, FA_1,\ldots, FA_k, \mu]$ in a FOODB model, *ID* is a unique object identity attribute, FA_i is an attribute, and $\mu \in [0, 1]$ denotes that a class instance may belong to the class with a membership degree of [0, 1]. Then the class name *FC* becomes the name of the element type, but the attributes cannot be transformed into element content description directly. The fuzzy class *FC* is transformed into the following fuzzy XML DTD elements:

<!ELEMENT *FC* (Dist)>
<!ATTLIST *FC ID* ID #REQUIRED>
<!ELEMENT Dist (Val+)>
<!ATTLIST Dist type (disjunctive|conjunctive)>
<!ELEMENT Val (FA_1?,…, FA_k?)>
<!ATTLIST Val Poss CDATA “1.0”>

Rule 4: For each class $FC = [ID, FA_1,\ldots, FUZZY\ FA_i, FA_k]$ in a FOODB model, *FUZZY* FA_i denotes that the attribute FA_i is a fuzzy attribute and it may take fuzzy value, and $1 \le i \le k$. In this case, the content types of the elements

corresponding to the fuzzy attributes are *Dist* instead of the element content description with content type #PCDATA.

For example, given a fuzzy class "*Young-Students* = [*YID*, *Name*, *FUZZY Age*, *FUZZY Height*]" in a FOODB model as mentioned in Sect. 2.5, firstly, an element type:

<!ELEMENT *Young-Students* (*Name*?, *Age*?, *Height*?)>
<!ATTLIST Young-Students *YID* IDREF #REQUIRED>

is created. Here the fuzzy class *Young-Students* in a FOODB model is transformed into an element type *Young-Students* in a fuzzy XML DTD with Rule 1, the object identity attribute *ID* in the fuzzy class is transformed into an attribute description of the element *Young-Students* and the other attributes are transformed into the element content description of the element *Young-Students* with Rule 2. Further, the content type of the crisp attribute "*Name*" is #PCDATA, i.e., we have:

<!ELEMENT *Name* (#PCDATA)>

But it is clear that the two attributes, "*Age*" and "*Height*", take fuzzy values. Therefore, they cannot be directly transformed into the element content description with content type #PCDATA. The content type of the fuzzy attribute "*Age*" is:

<!ELEMENT *age* (Dist)>
<!ELEMENT Dist (Val+)>
<!ATTLIST Dist type (disjunctive)>
<!ELEMENT Val (#PCDATA)>
<!ATTLIST Val Poss CDATA "1.0">

Finally, the fuzzy class "*Young-Students* = [*YID*, *Name*, *FUZZY Age*, *FUZZY Height*]" in a FOODB model is transformed into the following fuzzy XML DTD:

<!ELEMENT *Young-Students* (*Name*?, *Age*?, *Height*?)>
<!ATTLIST Young-Students *YID* IDREF #REQUIRED>
<!ELEMENT *Name* (#PCDATA)>
<!ELEMENT *Age* (Dist)>
<!ELEMENT Dist (Val+)>
<!ATTLIST Dist type (disjunctive)>
<!ELEMENT Val (#PCDATA)>
<!ATTLIST Val Poss CDATA "1.0">
<!ELEMENT *Height* (Dist)>
<!ELEMENT Dist (Val+)>
<!ATTLIST Dist type (disjunctive)>
<!ELEMENT Val (#PCDATA)>
<!ATTLIST Val Poss CDATA "1.0">

Rule 5: For the multiplicity specifications of attributes in a fuzzy class $FC = [\ldots, FA_i\ (m, n), \ldots]$ in a FOODB model, the optional multiplicity (m, n) for the attribute FA_i specifies that FA_i associates to each instance of FC at least m and most n instances of its type. The multiplicity is transformed into the cardinality

specifications (with specifies ?, *, +) used for element content construction, e.g., [0, 1] is transformed into ?.

5.4.2 Extraction of Fuzzy Relationships

The relationships describe connections among class instances. A relationship is a more general relation than the inheritance of classes. In general, the relationships are similar to the associations mentioned in fuzzy UML models in Sect. 5.2.3. In the following, we transform the relationships of classes in a FOODB model into fuzzy XML using the approach of transforming the associations in fuzzy UML models introduced in Sect. 5.2.3.

Rule 6: For each relationship *FR* between two classes FC_1 and FC_2 in a FOODB model such that FR = [..., FC_1: FC_2,...] in a FOODB model, $FCID_1$ and $FCID_2$ are the unique object identity attributes of FC_1 and FC_2, respectively, and the fuzzy classes FC_1 and FC_2 may have attribute FA_i and μ as mentioned in the Rules 3 and 4 above. Then the following fuzzy XML DTD element types are created:

- the class names FC_1 and FC_2 become the names of the element types, and their transformations are carried out according to the transformations of fuzzy classes introduced in the Rules 3 and 4 above:

  <!ELEMENT FC_1 (...)>
  <!ELEMENT FC_2 (...)>

- each transformed element must be augmented by a #REQUIRED IDREF attribute (ISIS XML/EDI Project, 2001), which is an artificial one and from another class involved in the relationship, i.e., we have:

  <!ELEMENT FC_1 (...)>
    <!ATTLIST FC_1 $FCID_2$ IDREF #REQUIRED>
  <!ELEMENT FC_2 (...)>
    <!ATTLIST FC_2 $FCID_1$ IDREF #REQUIRED>

Here we provide a transformation example. Given a relationship "*take-course* = [*Young-Students*: *Course*]", where *Young-Students* = [*YID*, *Name*, *FUZZY Age*, *FUZZY Height*] and *Course* = [*CID*, *Cname*, μ], by the Rule 6, the created fuzzy XML DTD is as follows:

```
<!ELEMENT Young-Students (Name?, Age?, Height?)>
    <!ATTLIST Young-Students YID IDREF #REQUIRED>
    <!ATTLIST Young-Students CID IDREF #REQUIRED>
<!ELEMENT Name (#PCDATA)>
<!ELEMENT Age (Dist)>
<!ELEMENT Dist (Val+)>
```

<!ATTLIST Dist type (disjunctive)>
<!ELEMENT Val (#PCDATA)>
<!ATTLIST Val Poss CDATA "1.0">
<!ELEMENT *Height* (Dist)>
<!ELEMENT Dist (Val+)>
<!ATTLIST Dist type (disjunctive)>
<!ELEMENT Val (#PCDATA)>
<!ATTLIST Val Poss CDATA "1.0">
<!ELEMENT *Course* (Dist)>
<!ATTLIST *Course CID* ID #REQUIRED>
<!ATTLIST *Course YID* IDREF #REQUIRED>
<!ELEMENT Dist (Val+)>
<!ATTLIST Dist type (disjunctive)>
<!ELEMENT Val (*Cname*?)>
<!ATTLIST Val Poss CDATA "1.0">
<!ELEMENT *Cname* (#PCDATA)>

5.4.3 Extraction of Fuzzy Inheritance Hierarchies

In FOODB models, classes may be fuzzy. A class produced from a fuzzy class must be fuzzy. If the former is still called subclass and the later superclass, the subclass/superclass relationship is fuzzy. The following rule can transform the fuzzy inheritance hierarchies into the fuzzy XML DTD element types.

Rule 7: For each fuzzy inheritance hierarchy such that $FC = [FC_1, \ldots, FC_n]$ in a FOODB model, $FC_1, \ldots, FC_n$ are the subclasses and FC is the superclass. Then the following fuzzy XML DTD element types are created:

- the class names FC, $FC_1, \ldots,$ and FC_n become the names of the element types, and the element type originating from the superclass is called a *superelement* and the element type originating from a subclass is called a *subelement*. The transformations of these classes are carried out according to the transformations of fuzzy classes introduced in the Rules 3 and 4 above:

<!ELEMENT FC (…)>
<!ELEMENT FC_1 (…)>
…
<!ELEMENT FC_n (…)>

- the superelement and each subelement must be associated with ID #REQUIRED and IDREF #REQUIRED, respectively, i.e., we have:

<!ELEMENT FC (…)>
<!ATTLIST FC $FCID$ ID #REQUIRED>
<!ELEMENT FC_1 (…)>

```
    <!ATTLIST FC1 FCIDref IDREF #REQUIRED>
...
  <!ELEMENT FCn (...)>
    <!ATTLIST FCn FCIDref IDREF #REQUIRED>
```

In the following we present a transformation example. Given a fuzzy inheritance hierarchy such that *Students* = [*Young-Students*, *Old-Students*] in a FOODB model, where *Students* = [*SID*, *Dept*], *Young-Students* = [*YID*, *Name*, *FUZZY Age*, *FUZZY Height*], and *Old-Students* = [*OID*, *Grade*], by the Rule 7, the created fuzzy XML DTD is as follows:

```
<!ELEMENT Students (Dept?)>
  <!ATTLIST Students SID ID #REQUIRED>
<!ELEMENT Dept (#PCDATA)>
<!ELEMENT Young-Students (Name?, Age?, Height?)>
  <!ATTLIST Young-Students YID IDREF #REQUIRED>
<!ELEMENT Name (#PCDATA)>
<!ELEMENT Age (Dist)>
<!ELEMENT Dist (Val+)>
  <!ATTLIST Dist type (disjunctive)>
<!ELEMENT Val (#PCDATA)>
  <!ATTLIST Val Poss CDATA "1.0">
<!ELEMENT Height (Dist)>
<!ELEMENT Dist (Val+)>
  <!ATTLIST Dist type (disjunctive)>
<!ELEMENT Val (#PCDATA)>
  <!ATTLIST Val Poss CDATA "1.0">
<!ELEMENT Old-Students (Grade?)>
  <!ATTLIST Old-Students OID IDREF #REQUIRED>
<!ELEMENT Grade (#PCDATA)>
```

5.5 Summary

With the development of various fuzzy database models and the wide utilization of the Web, there is an increasing need to effectively publish fuzzy data in fuzzy database models as fuzzy XML documents for Web-based applications. In this chapter, we considered the extraction of fuzzy XML from several typical fuzzy database models including fuzzy UML data models, fuzzy relational database models, and fuzzy object-oriented database models, provided a set of rules which successfully handles the extraction process, and also introduced the extraction approaches in detail by illustrating transformation examples.

The transformations from the fuzzy database models to the fuzzy XML models lay a theoretical foundation for establishing the overall management system of fuzzy XML data. Moreover, as being introduced in the following chapter, how to reengineer fuzzy XML data into fuzzy database models is considered very interesting to satisfy the needs of storing fuzzy XML data in fuzzy databases.

References

Conrad R, Scheffner D, Freytag JC (2000) XML conceptual modeling using UML. In: Proceedings of the 19th international conference on conceptual modeling, pp 558–571

Fernández M, Tan WC, Suciu D (2000) SilkRoute: trading between relations and XML. In: Proceedings of the 9th international World Wide Web conference on computer networks: the international journal of computer and telecommunications networking, pp 723–745

Fong J, Cheung SK (2005) Translating relational schema into XML schema definition with data semantic preservation and XSD graph. Inf Softw Technol 47:437–462

ISIS XML/EDI Project (2001) Mapping from UML generalised message descriptions to XML DTDs. http://palvelut.tieke.fi/edi/isis-xmledi/d2/UmlToDtdMapping05.doc

Lee DW, Mani M, Chiu F, Chu WW (2001) Nesting-based relational-to-XML schema translation. In: Proceedings of the 4th international workshop on the web and databases. California, USA, pp 61–66

Liu J, Ma ZM (2013) Formal transformation from fuzzy object-oriented databases to fuzzy XML. Appl Intell. doi:10.1007/s10489-013-0438-4

Liu C, Vincent MW, Liu J (2006) Constraint preserving transformation from relational schema to XML schema. World Wide Web 9(1):93–110

Lo A, Ozyer T, Kianmehr K, Alhajj R (2010) VIREX and VRXQuery: interactive approach for visual querying of relational databases to produce XML. J Intell Inf Syst 35(1):21–49

Ma ZM, Yan L (2007) Fuzzy XML data modeling with the UML and relational data models. Data Knowl Eng 63(3):970–994

Naser T, Alhajj R, Ridley MJ (2009) Two-way mapping between object-oriented databases and XML. Informatica 33(3):297–308

Wang C, Lo A, Alhajj R, Barker K (2005) Novel approach for reengineering relational databases into XML. In: Proceedings of international conference on data engineering, pp 1284–1289

Yan L, Ma ZM, Liu J, Zhang F (2011) XML modeling of fuzzy data with relational databases. Chin J Comput 34(2):291–303

Chapter 6
Reengineering Fuzzy XML into Fuzzy Database Models

Abstract Since the simplicity and flexibility of eXtensible Markup Language (XML), it has become the lingua franca for data exchange on the Web. Also, in order to deal with imprecise and uncertain information in many real-world applications, fuzzy XML has been extensively investigated as introduced in Chap. 3. However, XML brings some limitations, e.g., it may be difficult to store various data in a semantics way because of the semi-structured characteristic of XML. As we have known, fuzzy databases such as fuzzy relational databases and fuzzy object-oriented databases can store a large set of semantic information. Therefore, there is an increasing need to reengineer fuzzy XML into fuzzy database models, which may satisfy the needs of storing fuzzy XML data in fuzzy databases. In this chapter, we focus on reengineering fuzzy XML into fuzzy database models, including fuzzy UML data models, fuzzy relational database models, and fuzzy object-oriented database models.

6.1 Introduction

With the prompt development of the Internet, the requirement of managing information based on the Web has attracted much attention both from academia and industry. The eXtensible Markup Language (XML) is widely regarded as the next step in the evolution of the World Wide Web, and has been the de-facto standard. It aims at enhancing content on the World Wide Web. XML and related standards are flexible that allow the easy development of applications which exchange data over the web such as e-commerce (EC) and supply chain management (SCM). However, this flexibility makes it challenging to develop an XML management system. To manage XML data, it is necessary to integrate XML and databases (Bertino and Catania 2001). As we have known, most of the data nowadays reside in databases, it is important to automate the process of storing or converting XML documents in databases; this will help in better analysis of the data using the efficient querying facilities of the existing databases (Naser et al. 2008). Moreover, many applications

L. Yan et al., *Fuzzy XML Data Management*, Studies in Fuzziness and Soft Computing 311, DOI: 10.1007/978-3-642-44899-7_6,

or softwares do not always understand XML and have fixed relational interfaces. Therefore, you might sometimes find it useful to shred all or some of the data values of an incoming XML document into rows and columns of relational tables. Also, the additional motivation to carry out the conversions is the fact that it is necessary to facilitate platform independent exchange of the content of databases and the need to store XML in databases (Naser et al. 2008, 2009). Accordingly, many approaches and tools have been proposed to convert or store XML to databases such as UML, relational and object-oriented databases. Conceptual data modeling of XML document schema (Conrad et al. 2000; Elmasri et al. 2005; Mani et al. 2001; Psaila 2000) and XML Schema (Bernauer et al. 2004) have been studied in the recent past. In Conrad et al. (2000), for example, UML was used for designing XML DTD (document type definition). Since the conceptual data models with powerful data abstraction contain clear and rich semantics and do not have data type limitation, the information integration based on conceptual data models is more advantageous than the information integration based on logical database models. For the integration of XML data, first XML document can be transformed into conceptual models (dos Santos Mello and Heuser 2001), and then the transformed conceptual models are integrated together. Moreover, from XML to relational database, some conversion approaches were developed. Deutsch et al. (1998) made the first attempts to store XML in relational databases, and used a data mining technique to find a DTD whose support exceeds the pre-defined threshold and using the DTD, converted XML documents to relational databases. How to store and query XML using relational database techniques were investigated in Florescu and Kossmann (1999), Shanmugasundaram et al. (1999), Dayen (2001), Bourret (1999). Lee and Chu (2001) developed an approach, where the hidden semantic constraints in DTDs are systematically found and translated into relational formats. Also, Lee et al. (2003) discussed the schema conversion methods between XML and relational models, and gave a brief overview of such techniques. As the mapping from XML to object-oriented databases is concerned, the work described in Chung et al. (2001) generates an object-oriented database schema from DTDs, stores it into the object-oriented database and processes XML queries; it mainly concentrates on representing the semi-structural part of XML data by inheritance. Also, other efforts on reengineering XML data into object-oriented databases can be found in Naser et al. (2008, 2009).

However, in many practical applications, it is difficult to state all information with one hundred percent certainty, and information imprecision and uncertainty often exist in these applications. Therefore, in order to handle the imprecise and uncertain information, fuzzy data modeling has been widely investigated in various database models, including fuzzy conceptual data models (fuzzy ER/EER model, fuzzy UML model, and etc.) and fuzzy logical database models (fuzzy relational database model, fuzzy object-oriented database model, and etc.) as introduced in Chap. 2. Moreover, information fuzziness has also been investigated in the context of e-commence (EC) and supply chain management (SCM) (Petrovic et al. 1999; Yager 2000; Yager and Pasi 2001). It is shown that fuzzy set theory is very useful in Web-based business intelligence. Unfortunately, although it is the current standard for data representation and exchange over the Web, XML is not able to represent and process imprecise and uncertain data. In this case, many approaches have been developed to introduce fuzziness in XML documents as introduced in Chap. 3. But it can be found that the study of reengineering fuzzy XML in fuzzy database models has received little attention. Fuzzy databases such as fuzzy relational databases and fuzzy object-oriented databases can store a large set of semantic information. Therefore, reengineering fuzzy XML into fuzzy database models may satisfy the needs of storing fuzzy XML data in fuzzy databases. Hollander and van Keulen (2010) investigated how to store and query probabilistic XML using a probabilistic relational database. Liu et al. (2013) presented an approach for reengineering fuzzy XML into fuzzy object-oriented database models. Ma and Yan (2007) investigated the formal mapping from the fuzzy XML model to the fuzzy relational databases, where a fuzzy DTD tree is created from the hierarchical XML DTD, and then the formal mapping from the fuzzy DTD tree to the fuzzy relational schema is developed. Yan et al. (2009) investigated the formal conversions from the fuzzy XML model to the fuzzy UML model. In this chapter, we introduce the issue on reengineering fuzzy XML into fuzzy database models, including fuzzy UML data model, fuzzy relational database model, and fuzzy object-oriented database model.

6.2 Reengineering Fuzzy XML into Fuzzy UML Data Models

XML lacks sufficient power in modeling real-world data and their complex interrelationships in semantics. So it is necessary to use other methods to describe data paradigms and develop a true conceptual data model, and then transform this model into an XML encoded format. Conceptual data models are able to capture and represent rich and complex semantics at a high abstract level and can be used for conceptual design of databases as well as XML. Therefore, in this section, based on the fuzzy UML data model and the fuzzy XML DTD model, we present the formal approach to mapping the fuzzy DTD model into the fuzzy UML data model.

To transform the fuzzy XML DTD to the fuzzy UML data model, we need a fuzzy DTD tree created from the hierarchical fuzzy XML DTD. We first construct a DTD tree through parsing the given fuzzy DTD model and then map the DTD tree into the fuzzy UML data model.

Generally, nodes in a DTD tree are element and attributes, in which each element appears exactly once in the graph, while attributes appear as many time as they appear in the DTD. The element nodes can be further classified into two kinds, that is, *leaf element nodes* and *nonleaf element nodes*. Thus in the DTD tree, we have three kinds of nodes, which are *attribute nodes*, *leaf element nodes* and *nonleaf element nodes*. There exists a special nonleaf element node in the DTD tree, namely the *root node*. We also need to identify such attribute nodes that the corresponding attributes are associated with ID #REQUIRED or IDREF #REQUIRED in DTD. We call these attribute nodes *key attribute nodes*. In addition, different to the classical DTD tree, the fuzzy DTD tree contains some new attribute and element types, which are attribute Poss and element Val and Dist as mentioned in Sect. 3.3.

A fuzzy DTD tree can be constructed when parsing the given fuzzy DTD following the ensuring processing:

(a) Take the first nonleaf element *r* of the given hierarchical DTD and create a DTD tree rooted at *r*. *r*'s children come from the attributes and elements connecting with *r*. Here, the key attribute(s) should become the primary key attribute(s) of the created DTD tree.
(b) Take the nonleaf element *s* of *r*'s child in the given hierarchical DTD and create a DTD sub-tree rooted at *s*. We apply the processing given in (a) to treat the *s*'s children.
(c) For other nonleaf elements in the given hierarchical DTD, apply the same processing given in (b) until all nonleaf elements are transformed.
(d) For all the generated sub-trees, we stitch them together and construct the fuzzy DTD tree.

According to Ma and Yan (2007), in the fuzzy DTD tree, in addition to (key) attribute nodes, leaf element nodes, and nonleaf element nodes, there are three special nodes, which are *Poss attribute nodes*, *Val element nodes*, and *Dist element nodes*. The *Dist* element nodes created from *Disk* elements are used to indicate the type of a possibility distribution, being *disjunctive* or *conjunctive*. In addition, each Dist element node has a Val element node as its child node, and a nonleaf element node as its parent node. Also, we can identify four kinds of Val element nodes as follows:

(a) They do not have any child node except the Poss attribute nodes (*type-1*).
(b) They only have leaf element nodes as their child nodes except the Poss attribute nodes (*type-2*).
(c) They only have nonleaf element nodes as their child nodes except the Poss attribute nodes (*type-3*).
(d) They have leaf element nodes as well as nonleaf element nodes as their child nodes except the Poss attribute nodes (*type-4*).

In the transformation of the fuzzy DTD tree to the fuzzy UML data model, the Poss attribute nodes, Val element nodes and Dist element nodes in the fuzzy DTD tree do not take part in composing the created fuzzy UML data model and only determine the model of the created fuzzy UML data model.

We take the root node of the given fuzzy DTD tree and create a class. The attributes of this class first come from the attribute nodes and leaf element nodes connecting with the root node. Here, the key attribute node(s) should become the primary key attribute(s) of the created class. Then it is needed to determine if the root node has any Val element nodes or Dist element nodes as its child nodes. If yes, we need to further determine the type of each Val element node (we can ignore Dist element nodes because each Dist element node must have a Val element node as its child node only). Note that it is impossible that the Val element nodes of *type-1* arise in the root node.

1. If it is the Val element node of *type-2*, all of the leaf element nodes connecting with the Val element node become the attributes of the created class. An additional attribute μ is also added into the created class.
2. If it is the Val element node of *type-3*, and the Val element's children except the Poss attribute nodes, namely nonleaf element nodes have nonleaf element nodes as their child, an additional attribute μ is added into the created class, and we leave the nonleaf element nodes for further treatment being discussed below.
3. If it is the Val element node of *type-4*, we do the same thing as (2) for the leaf element nodes and the nonleaf element nodes that only have leaf element nodes as their child. And we leave the nonleaf element nodes that have nonleaf element as their child for further treatment being discussed below.

For each nonleaf element node that has nonleaf element as their child connecting with the root node, we create a separate class. Its attributes come from the attribute nodes and leaf element nodes connecting with this nonleaf element node, and its primary key attribute(s) should come from the key attribute node(s). Furthermore, it is needed to determine if this nonleaf element node has any Val element nodes or Dist element nodes as its child nodes, and further identify the type of these nodes, if any. We apply the processing given in (1)–(3) above to treat the Val element nodes of *type-2*, *type-3*, and *type-4*. For the Val element nodes of *type-1*, each of them should become an attribute of the class created from the parent node of the current nonleaf element. Note that this attribute is one that may take fuzzy values. For other nonleaf element nodes in the fuzzy DTD tree, we continue to apply the same processing above until all nonleaf element nodes are transformed.

Note that, for some elements which are not Dist, Val and leaf element, we may selectively use separate classes, in which the key attributes of these elements become the primary attributes of the separate classes. Considering easy maintenance of data in the entity type, we choose a class for one nonleaf element that only has leaf element nodes as their child.

Now we use an example to illustrate the transformation of the fuzzy XML DTD to the fuzzy UML data model. Figure 6.1 shows a brief fuzzy DTD tree created basically from the hierarchical fuzzy XML DTD given in Chap. 3. In this fuzzy DTD tree, the Dist element nodes created from Dist elements are used to indicate the type of a possibility distribution, being disjunctive or conjunctive. In addition, each Dist element node has a Val element node as its child node, and a nonleaf element node as its parent node.

During the transformation, we first create a class "university", in which *Uname* is the primary key attribute of the class, and the leaf element node *address*

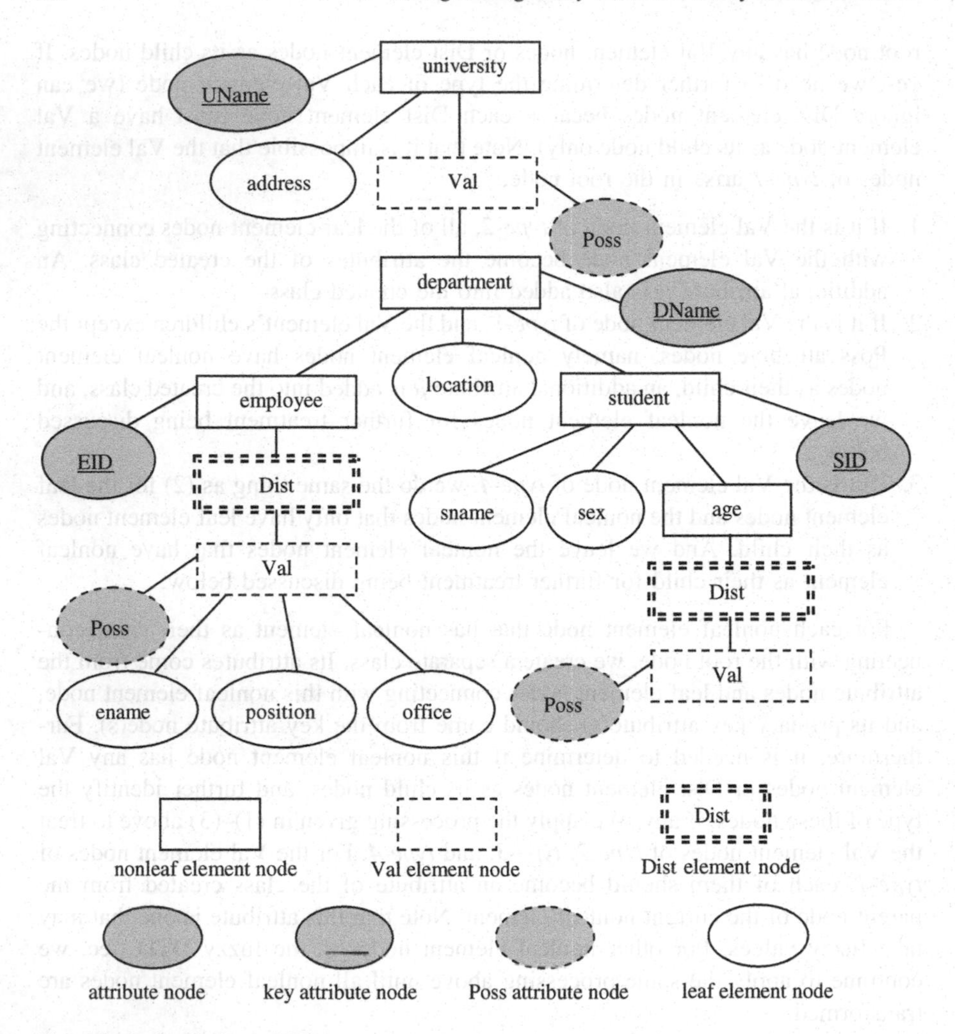

Fig. 6.1 A simple fuzzy DTD tree

connecting with the university is a single attribute of the class. For Val element, according to (2) above, we create an additional attribute μ in the class. For the nonleaf element "department", we find that it has two children *employee* and *student*, and both of them are nonleaf element. Then we create three classes "department", "employee" and "student". In particular, class "employee" contains five attributes *EID*, *ename*, *position*, *office* and an additional attribute μ, and class "student" contains four attributes *SID*, *sname*, *sex* and *age*, in which attribute *age* is one that may take fuzzy values.

After transformation, the fuzzy DTD tree in Fig. 6.1 is mapped into the fuzzy UML data model shown in Fig. 6.2.

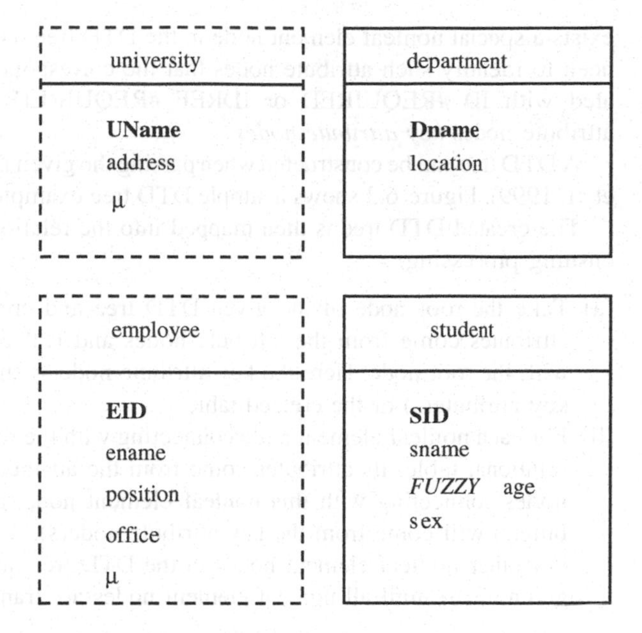

Fig. 6.2 A simple fuzzy UML model reengineering from the fuzzy DTD tree in Fig. 6.1

6.3 Reengineering Fuzzy XML into Fuzzy Relational Database Models

To reengineer fuzzy XML into fuzzy relational database model, Ma and Yan (2007) investigated the formal mapping from the fuzzy XML model to the fuzzy relational database. Being similar to the mapping from the fuzzy XML model to the fuzzy UML data model as introduced in Sect. 6.2, when reengineer fuzzy XML into fuzzy relational database model, a fuzzy DTD tree is also needed to be created from the hierarchical XML DTD, and then the formal mapping from the fuzzy DTD tree to the fuzzy relational schema is developed.

6.3.1 DTD Tree and Mapping to the Relational Database Schema

The hierarchical XML and the flat relational data models are not fully compliant so the transformation is not a straightforward task. As introduced in Sect. 6.2, generally, a DTD tree can be created from the hierarchical XML DTD. Its nodes are elements and attributes, in which each element appears exactly once in the graph, while attributes appear as many times as they appear in the DTD. The element nodes can be further classified into two kinds: leaf element nodes and nonleaf element nodes. So in the DTD tree, we have three kinds of nodes, which are *attribute nodes*, *leaf element nodes,* and *nonleaf element nodes*. Note that there exists a special nonleaf element node in the DTD tree, i.e., the *root node*. We also need to identify such attribute nodes that the corresponding attributes are associated with ID #REQUIRED or IDREF #REQUIRED in DTD. We call these attribute nodes *key attribute nodes*.

A DTD tree can be constructed when parsing the given DTD (Shanmugasundaram et al. 1999). Figure 6.3 shows a simple DTD tree example.

The created DTD tree is then mapped into the relational schema following the ensuing processing:

(a) Take the root node of the given DTD tree and create a relational table. Its attributes come from the attribute nodes and leaf element nodes connecting with the root node. Here the key attribute node(s) should become the primary key attribute(s) of the created table.
(b) For each nonleaf element node connecting with the root node, create a separate relational table. Its attributes come from the attribute nodes and leaf element nodes connecting with this nonleaf element node, and its primary key attribute(s) will come from the key attribute node(s).
(c) For other nonleaf element nodes in the DTD tree, apply the same processing given in (b) until all nonleaf element nodes are transformed.

Note that there may be cycles in DTD and element declarations that are referenced from more than one element declaration as element contents. Then we

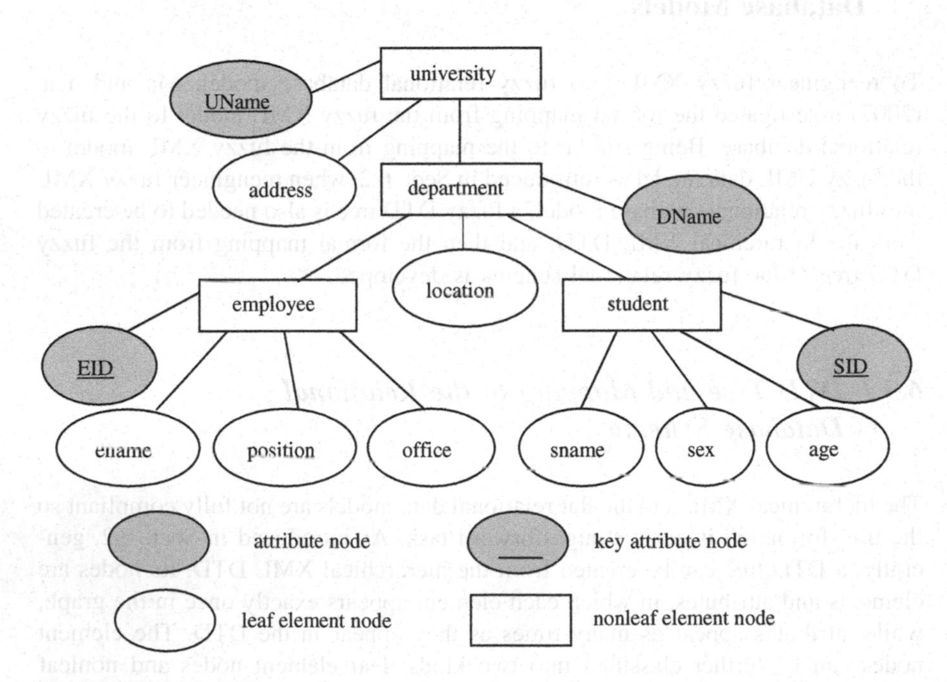

Fig. 6.3 A simple DTD tree

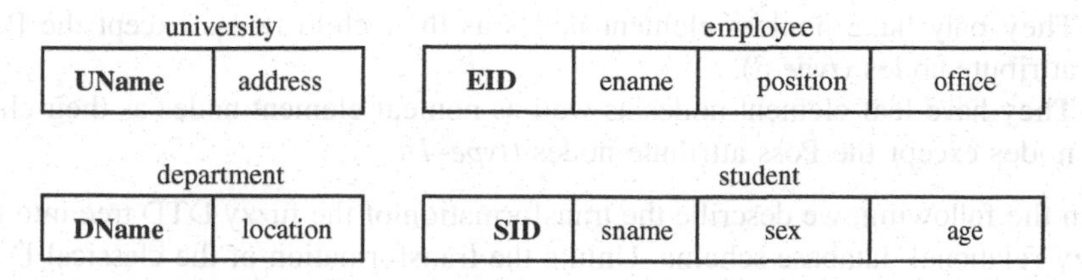

Fig. 6.4 The relational schemas created by the DTD tree in Fig. 6.3

need to link a created relational table to its parent relational table through the parent table's primary key.

The DTD tree in Fig. 6.3 is mapped into the relational schemas shown in Fig. 6.4.

6.3.2 Mapping the Fuzzy XML Model into the Fuzzy Relational Database Model

Generally speaking, the fuzzy XML DTD presented in Chap. 3 can be transformed into the fuzzy relational database schema using a similar processing as given above under a classical environment. That is, we first construct a DTD tree through parsing the given fuzzy DTD, and then map the DTD tree into the fuzzy relational database schema. However, the DTD tree here, called the *fuzzy DTD tree*, is clearly different from the classical DTD tree above because the fuzzy DTD contains new attribute and element types, which are attribute Poss and elements Val and Dist. As a result, the transformation of the fuzzy DTD tree to the fuzzy relational database schema is also different from the transformation of the classical DTD tree to the classical relational database schema.

In the fuzzy DTD tree, in addition to (key) attribute nodes, leaf element nodes, and nonleaf element nodes, there are three special nodes, which are *Poss attribute nodes*, *Val element nodes*, and *Dist element nodes*. Figure 6.1 shows a simple fuzzy DTD tree that basically comes from the fuzzy DTD given in Chap. 3. In this fuzzy DTD tree, the Dist element nodes created from Disk elements are used to indicate the type of a possibility distribution, being *disjunctive* or *conjunctive*. In addition, each Dist element node has a Val element node as its child node, and a nonleaf element node as its parent node.

Here, we briefly recall the four kinds of Val element nodes in Fig. 6.1 as mentioned in Sect. 6.2:

(a) They do not have any child node except the Poss attribute nodes (*type-1*).
(b) They only have leaf element nodes as their child nodes except the Poss attribute nodes (*type-2*).

(c) They only have nonleaf element nodes as their child nodes except the Poss attribute nodes (*type-3*).
(d) They have leaf element nodes as well as nonleaf element nodes as their child nodes except the Poss attribute nodes (*type-4*).

In the following, we describe the transformation of the fuzzy DTD tree into the fuzzy relational database schema. Unlike the transformation of the classical DTD tree to the relational database schema, in the transformation of the fuzzy DTD tree to the fuzzy relational model, the Poss attribute nodes, Val element nodes, and Dist element nodes in the fuzzy DTD tree do not take part in composing the created relational schema and only determine the model of the created fuzzy relational databases. Being similar to the process in Sect. 6.2, we have the following process:

(a) Take the root node of the given fuzzy DTD tree and create a relational table. Its attributes first come from the attribute nodes and leaf element nodes connecting with the root node. Here, the key attribute node(s) should become the primary key attribute(s) of the created table. Then determine if the root node has any Val element nodes or Dist element nodes as its child nodes. If yes, we need to further determine the type of each Val element node (we also can ignore Dist element nodes because each Dist element node must have a Val element node as its child node only).
    (i) If it is the Val element node of *type-3*, only an additional attribute is added into the created relational table, representing the possibility degree of the tuples.
    (ii) If it is the Val element node of *type-2*, all of the leaf element nodes connecting with the Val element node become the attributes of the created relational table. An additional attribute is also added into the created relational table, representing the possibility degree of the tuples.
    (iii) If it is the Val element node of *type-4*, we leave the nonleaf element nodes for further treatment in (b) and do the same thing as (ii) for the leaf element nodes.

It is impossible that the Val element nodes of *type-1* arise in the root node.

(b) For each nonleaf element node connecting with the root node, create a separate relational table. Its attributes come from the attribute nodes and leaf element nodes connecting with this nonleaf element node, and its primary key attribute(s) will come from the key attribute node(s). Furthermore, determine if this nonleaf element node has any Val element nodes or Dist element nodes as its child nodes, and identify the type of these nodes, if any. We still apply the processing given in (i)–(iii) of (a) to treat the Val element nodes of *type-2*, *type-3* and *type-4*. For the Val element nodes of *type-1*, each of them should become an attribute of another relational table created from the parent node of the current nonleaf element. Note that this attribute is one that may take fuzzy values.
(c) For other nonleaf element nodes in the fuzzy DTD tree, apply the same processing given in (b) until all nonleaf element nodes are transformed.

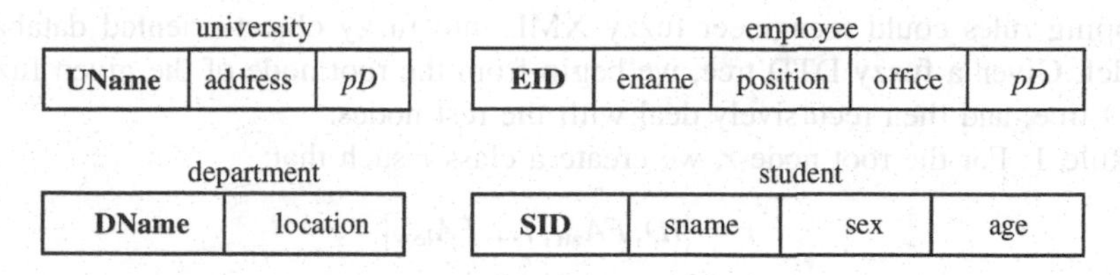

Fig. 6.5 The fuzzy relational schemas created by the fuzzy DTD tree in Fig. 6.1

Note that we could alternatively use a single relational table for some elements which are not the Dist, Val, and leaf ones. In this case, all key attribute nodes from these elements should become the primary key attributes of the single table. Considering the high independence and easy maintenance of data in the table as well as the normalization of relational schema, we generally choose a relational table for one nonleaf element.

The fuzzy DTD tree in Fig. 6.1 is mapped into the fuzzy relational schemas shown in Fig. 6.5, in which attribute "age" is one that may take fuzzy values.

6.4 Reengineering Fuzzy XML into Fuzzy Object-Oriented Database Models

The classical relational database model and its fuzzy extension do not satisfy the need of modeling complex objects with imprecision and uncertainty. In order to model uncertain data and complex-valued attributes as well as complex relationships among objects, current efforts have concentrated on the fuzzy object-oriented databases as introduced in Chap. 2. Therefore, reengineering fuzzy XML into fuzzy object-oriented database model may satisfy the needs of storing fuzzy XML data in fuzzy databases and help to the interoperability between fuzzy object-oriented database model and fuzzy XML. Based on the similar idea in Sects. 6.2 and 6.3, in the following, we introduce how to reengineer fuzzy XML into fuzzy object-oriented database model, and provide a set of rules for mapping fuzzy XML into fuzzy object-oriented database model.

As mentioned in Sects. 6.2 and 6.3, a fuzzy XML DTD can be transformed into a fuzzy DTD tree. In the fuzzy DTD tree, in addition to (key) attribute nodes, leaf element nodes, and nonleaf element nodes, there are three special nodes, which are *Poss*, *Val*, and *Dist*. The *Dist* element nodes created from *Dist* elements are used to indicate the type of a possibility distribution, being *disjunctive* or *conjunctive*. In addition, each Dist element node has a Val element node as its child node, and a nonleaf element node as its parent node. In the transformation of the fuzzy DTD tree to the fuzzy object-oriented database model, the Poss attribute nodes, Val element nodes and Dist element nodes in the fuzzy DTD tree do not take part in composing the mapped fuzzy object-oriented database model and only determine the model of the created fuzzy object-oriented database model. The following mapping rules could reengineer fuzzy XML into fuzzy object-oriented database model. Given a fuzzy DTD tree, we begin from the root node of the given fuzzy DTD tree, and then recursively deal with the rest nodes.

Rule 1: For the root node *r*, we create a class *r* such that:

$$r = [ID, FA_{\text{att}}, \ldots, FA_{\text{leaf}}]$$

Here *ID* is the unique object identity of the class *r* from the key attribute node of the root node in the fuzzy DTD tree; FA_{att} is the attribute of the class *r* from the attribute nodes; and FA_{leaf} is also the attribute of the class *r* from the leaf element nodes connecting with the root node.

Then we further check if the root node *r* has any Val element nodes or Dist element nodes as its child nodes. If yes, the later Rules 3–6 will be used.

Rule 2: For each nonleaf element node *N* connecting with the root node, we create a class *N* such that:

$$N = [ID, FA_{\text{att}}, \ldots, FA_{\text{leaf}}]$$

Here *ID* is the unique object identity of the class *N* from the key attribute node of the nonleaf element node; FA_{att} and FA_{leaf} are the attributes of the class *N* from the attribute nodes and the leaf element nodes connecting with the nonleaf element node.

Specially, if the nonleaf element node only includes a Dist element node, and the Dist element node only includes a Val element node which does not have any child node except the Poss attribute node, then we do not create the class *N* for this nonleaf element node as being introduced in the following Rule 3.

Also, we need to check if this nonleaf element node has any Val element nodes or Dist element nodes as its child nodes, and then apply the following rules to these nodes. Here, being similar to the approaches introduced in Sects. 6.2 and 6.3, we also ignore Dist element nodes because each Dist element node has a Val element node as its child node only.

Rule 3: For each Val element node which does not have any child node except the Poss attribute nodes, the nonleaf element node connecting with the Val element node via a Dist element node becomes a fuzzy attribute, which is the attribute of the parent node of the nonleaf element node, and a keyword *FUZZY* followed by the attribute denotes that it is a fuzzy attribute.

For example (the example is a fragment of the fuzzy DTD D_1 in Fig. 3.2 in Chap. 3):

```
<!ELEMENT student (age?)>
  <!ATTLIST student SID IDREF #REQUIRED>
<!ELEMENT age (Dist)>
<!ELEMENT Dist (Val+)>
  <!ATTLIST Dist type (disjunctive)>
<!ELEMENT Val (#PCDATA)>
  <!ATTLIST Val Poss CDATA "1.0">
```

In the example, we note that the nonleaf element node *age* connects with the Val element node via a Dist element node. According to Rule 3, the fuzzy DTD is mapped into the class in a fuzzy object-oriented database model as follows:

Student = [SID, FUZZY age],

Here *student* is a created class, *SID* is its unique object identity, and *FUZZY age* is a fuzzy attribute of the parent node *student* of the nonleaf element node *age* and it may take fuzzy value which represents possibility distributions.

Rule 4: For each Val element node which only has leaf element nodes as its child nodes except the Poss attribute nodes, all of the leaf element nodes connecting with the Val element node become the attributes of the created class. An additional attribute μ is added into the created class.

For example:

```
<!ELEMENT employee (Dist)>
  <!ATTLIST employee FID IDREF #REQUIRED>
<!ELEMENT Val (fname?, position?, office?, course?)>
  <!ATTLIST Val Poss CDATA "1.0">
<!ELEMENT fname (#PCDATA)>
<!ELEMENT position (#PCDATA)>
<!ELEMENT office (#PCDATA)>
<!ELEMENT course (#PCDATA)>
```

In the fuzzy DTD, the Val element node has leaf element nodes *fname*, *position*, *office*, *course* as its child nodes, according to Rule 4, the fuzzy DTD is mapped into the class in a fuzzy object-oriented database model as follows:

employee = [FID, fname, position, office, course, μ],

Here *employee* is a created class, *FID* is its unique object identity, *fname*, *position*, *office*, *course* are its attributes, and $\mu \in [0, 1]$ denotes that a class instance may belong to the class with a membership degree of [0, 1].

Rule 5: For each Val element node which only has nonleaf element nodes as its child nodes except the Poss attribute nodes, an additional attribute μ is added into the created class.

For example:

```
<!ELEMENT university (Val+)>
  <!ATTLIST university UName IDREF #REQUIRED>
<!ELEMENT Val (department*)>
  <!ATTLIST Val Poss CDATA "1.0">
```

In the fuzzy DTD, the Val element node has nonleaf element node department as its child node, according to Rule 5, the fuzzy DTD is mapped into the class in a fuzzy object-oriented database model as follows:

university = [UName, μ],

Here *university* is a created class, *UName* is its unique object identity attribute, and $\mu \in [0, 1]$ denotes that a class instance may belong to the class with a membership degree of [0, 1].

Rule 6: For each Val element node which has leaf element nodes as well as nonleaf element nodes as their child nodes except the Poss attribute nodes, by jointly using the above rules, the nonleaf element nodes can be done as Rule 2 and the leaf element nodes can be handled as Rules 1–5.

Based on the rules above, each element node in the fuzzy DTD can be handled until all element nodes are transformed, and finally the fuzzy XML is reengineered into a fuzzy object-oriented database model.

6.5 Summary

With the prompt development of the Internet, the requirement of managing information based on the Web has attracted much attention both from academia and industry. XML is widely regarded as the next step in the evolution of the World Wide Web, and has been the de-facto standard. This creates a new set of data management requirements involving XML, such as the need to construct and store XML documents. On the other hand, fuzzy sets and possibility theory have been extensively applied to deal with information imprecision and uncertainty in the practical applications, and reengineering fuzzy XML into fuzzy database models is receiving more attention for managing fuzzy XML data. In this chapter, we proposed some approaches for reengineering fuzzy XML into fuzzy database models, including fuzzy UML data models, fuzzy relational database models, and fuzzy object-oriented database models, respectively.

The two-way mappings between the fuzzy database models to the fuzzy XML models pay an important role for establishing the overall management system of fuzzy XML data. Moreover, for processing fuzzy XML data intelligently, reasoning on the fuzzy XML data would help to check whether a fuzzy XML document conforms to a given document structure or two fuzzy XML documents are compatible, and also may improve the efficiency of query processing as will be introduced in the following chapter.

References

Bernauer M, Kappel G, Kramler G (2004) Representing XML schema in UML—A comparison of approaches. In: Proceedings of the 4th international conference on web engineering, Springer, Munich, pp 440–444

Bertino E, Catania B (2001) Integrating XML and databases. IEEE Internet Comput 5(4):84–88

Bourret R (1999) XML and databases. http://www.rpbourret.com/xml/XMLAndDatabases.htm

Chung TS, Park S, Han SY, Kim HJ (2001) Extracting object-oriented database schemas from XML DTDs using inheritance. In: Proceedings of the international conference on electronic commerce and web technologies, pp 49–59

Conrad R, Scheffner D, Freytag JC (2000) XML conceptual modeling using UML. In: Proceedings of the 19th international conference on conceptual modeling, Springer, Salt Lake City, pp 558–571

Dayen I (2001) Storing XML in relational databases. http://www.xml.com/pub/a/2001/06/20/databases.html

Deutsch A, Fernandez MF, Suciu D (1998) Storing semistructured data with STORED. In: Proceedings of the ACM SIGMOD, Philadelphia, PA, 431-442

dos Santos Mello R, Heuser CA (2001) A rule-based conversion of a DTD to a conceptual schema. In: Proceedings of the 20th international conference on conceptual modeling, Springer, Yokohama, pp 133–148

Elmasri R, Li Q, Fu J, Wu YC, Hojabri B, Ande S (2005) Conceptual Modeling for Customized XML Schemas. Data Knowl Eng 54:57–76

Florescu D, Kossmann D (1999) Storing and querying XML data using an RDBMS. IEEE Data Eng. 22:27–34

Hollander ES, and van Keulen M (2010) Storing and querying probabilistic XML using a probabilistic relational DBMS. In: Proceedings of the 4th international workshop on management of uncertain data (MUD 2010), pp 35–49

Lee D, Chu WW (2001) CPI: constraints-preserving inlining algorithm for mapping XML DTD to relational schema. J Data Knowl Eng 39:3–25

Lee D, Mani M, Chu WW (2003) Schema conversion methods between xml and relational models. In: Omelayenko B, Klein M (eds) Knowledge transformation for the semantic Web. IOS Press, Amsterdam, pp 1–17

Liu J, Ma ZM, Feng X (2013) Formal approach for reengineering fuzzy XML in fuzzy object-oriented databases. Appl Intell 38:541–555

Ma ZM, Yan L (2007) Fuzzy XML data modeling with the UML and relational data models. Data Knowl Eng 63:972–996

Mani M, Lee DW, Muntz RR (2001) Semantic data modeling using XML schemas. In: Proceedings of the 20th international conference on conceptual modeling, Springer, Yokohama, pp 149–163

Naser T, Alhajj R, Ridley MJ (2008) Reengineering XML into object-oriented database. In: Proceedings of the IEEE IRI, pp 1–6

Naser T, Alhajj R, Ridley MJ (2009) Two-way mapping between object-oriented databases and XML. Informatica 33(3):297–308

Petrovic D, Roy R, Petrovic R (1999) Supply chain modeling using fuzzy sets. Int J Prod Econ 59:443–453

Psaila G (2000) ERX: a data model for collections of XML documents. In: Proceedings of the 2000 ACM symposium on applied computing, ACM, Villa Olmo, pp 898–903

Shanmugasundaram J, Tufte K, He G, Zhang C, DeWitt D, Naughton J (1999) Relational databases for querying XML documents: limitations and opportunities. In: Proceedings of the VLDB, Edinburgh, Scotland, pp 302–314

Yager RR (2000) Targeted e-commerce marketing using fuzzy intelligent agents. IEEE Intell Systems 15(6):42–45

Yager RR, Pasi G (2001) Product category description for Web-shopping in e-commerce. Int J Intell Syst 16:1009–1021

Yan L, Ma ZM, Liu J, Yu G (2009) Formal conversion of fuzzy XML DTD to UML data model. J Chin Comput Syst 30(4):586–593

Chapter 7
Fuzzy XML Reasoning

Abstract XML has been the de-facto standard of information representation and exchange over the web. However, the real world is filled with imprecision and uncertainty. This creates a new set of data management requirements involving XML with imprecision and uncertainty, such as the need to reason on and query fuzzy XML documents and structures. Reasoning on XML with imprecision and uncertainty would help to check whether a fuzzy XML document conforms to a given document structure or two fuzzy XML documents are compatible, improve the precision and efficiency of query processing, etc. In particular, among several ways to approach knowledge representation and reasoning, Description Logics and ontologies are gaining privileged places in recent years. Therefore, in this chapter, we introduce how to reason on fuzzy XML with the knowledge representation formalisms fuzzy Description Logics and fuzzy ontologies.

7.1 Introduction

With the wide utilization of the Web and the availability of huge amounts of electronic data, *eXtensible Markup Language* (XML) has been the de-facto standard of information representation and exchange over the web (Bray et al. 1998). With the broad application of XML in real-world domains, this creates data management requirements involving XML, such as the need to represent, reason, integrate, and query XML documents and document structures. To this end, some work has investigated how to represent and reason on XML. The consistency/conformance problem of XML (i.e., whether an XML document conforms to a given document structure DTD) was studied in Arenas et al. (2002). Reasoning about equivalence in XML DTDs was introduced in Wood (1995), i.e., checking whether two DTDs define the same sets of XML document instances. The other efforts on how to represent and reason on XML have been carried out in Buneman et al. (2003), Baader et al. (2003), Wu et al. (2008), Cautis et al. (2007), Libkin and Sirangelo (2008), Toman and Weddell (2005), Calvanese et al. (1999).

L. Yan et al., *Fuzzy XML Data Management*, Studies in Fuzziness and Soft Computing 311, DOI: 10.1007/978-3-642-44899-7_7,

As mentioned in these researches, representing and reasoning on XML would help in several tasks related to XML, such as designing and integrating XML documents, checking reasoning tasks of XML, and providing query optimization.

In particular, among several ways to approach knowledge representation and reasoning, Description Logics and ontologies are gaining privileged places in recent years. *Description Logics* (*DLs*, for short), which are a family of knowledge representation languages that can be used to represent the knowledge of a domain in a structured and formally well-understood way, have been applied to various fields such as the Semantic Web, software engineering, and databases (Baader et al. 2003). *Ontologies* are a formal and explicit specification of a shared conceptualization and can enable semantic interoperability, and their logical foundation are *DLs*. Recently, based on the high expressive power and effective reasoning service, the *DLs* and ontologies have been extensively used to represent and reason on various data sources (e.g., ER model, UML model, etc.) (Baader et al. 2003; Borgida 1995; Staab and Studer 2004).

Also, it is not surprising that *DLs* and ontolgoies are particularly adept at representing and reasoning on XML. Reasoning on XML (e.g., conformance and equivalence) is a complex and time-consuming task by hand. Observe that the known algorithms for checking equivalence of two DTDs are doubly exponential time (Wood 1995). Therefore, if XML can be translated into *DL* knowledge bases or ontologies, the reasoning problems of XML may be reasoned through the reasoning mechanism of *DLs* and ontologies. To this end, a *DL* called $DLFD_{\text{reg}}$ was proposed in Toman and Weddell (2005) to reason on structural equality in XML. In (Wu et al. 2008), a *DL* approach to represent and reason on XML was developed. Moreover, how to establish the relationships between XML and *DLs*/ ontologies have also been investigated in Baader et al. (2003), Calvanese et al. (1998, 1999), Euzenat (2001), Drabent and Wilk (2006), Zhang et al. (2011a).

However, imprecise and uncertain information commonly exists in many real-world applications. In particular, the representation of fuzzy information with fuzzy set theory has been addressed several decades ago by Zadeh (1965). Over the years, the fuzzy set theory has been extensively introduced into databases, information systems, the Semantic Web, and so on (Bosc et al. 2005; Lukasiewicz and Straccia 2008; Galindo 2008). Also, the problems that emerge are how to handle fuzzy information within XML, and how to represent these non-crisp data within XML has received much attention in recent years. Regarding modeling imprecise and uncertain information in XML, lots of works have been done in modeling and querying imperfect XML data as have been introduced in the previous chapters. Accordingly, this also creates a new set of data management requirements involving XML with imprecision and uncertainty, such as the need to reason on and query fuzzy XML documents and document structures. Reasoning on XML with imprecision and uncertainty would help to check some reasoning tasks (e.g., whether a fuzzy XML document conforms a given document structure or two fuzzy XML documents are compatible), integrate fuzzy XML documents, improve the precision and efficiency of query processing, etc. Fortunately, currently, in order that *DLs* and ontologies can directly deal with imprecise and uncertain knowledge, a variety of fuzzy *DLs* and fuzzy ontologies have been proposed. The first effort on fuzzy DL was presented by Yen (1991). In his extension, explicit membership functions over a domain were used as well as membership manipulators , such as "very", in order to alter membership functions and define new concepts from already defined ones. A later approach was proposed in Tresp and Molitor (1998), where membership manipulators also appear. Typically, a fuzzy extension of the ALC language was considered in Straccia (1998), and the reasoning algorithm based on tableaux calculus was also provided in Straccia (2001). From them on, the approaches towards more expressive fuzzy *DLs* such as fuzzy ALCQ (Sánchez and Tettamanzi 2006), fuzzy ALCIQ (Stoilos et al. 2008) were presented. Furthermore, by extending the fuzzy ALC with transitive role axioms (*S*), inverse roles (*I*), role hierarchies (*H*) and number restrictions (*N*), Stoilos presented the fuzzy DL f-SHIN (Stoilos et al. 2005). A fuzzy extension of the corresponding description logic of the ontology description language OWL DL, called fuzzy SHOIN(D), was proposed in (Straccia 2005). A more expressive fuzzy DL called f-SROIQ(D) corresponding to a fuzzy extension of the OWL2 (an extension of OWL) was presented in (Bobillo et al. 2012). For a comprehensive review of fuzzy *DLs*, please refer to Lukasiewicz and Straccia (2008), Ma et al. (2013). Also, in order to represent and reason on fuzzy information in ontologies, some approaches have been developed to characterize or define fuzzy ontologies. Calegari and Ciucci (2007) integrated fuzzy logic in ontologies and developed a plug-in for the KAON Project in order to introduce fuzziness in ontologies. Lee et al. (2005) presented a four-layered fuzzy ontology and applied it to news summarization. Sanchez and Yamanoi (2006) introduced a fuzzy ontology structure from the aspects of lexicon and knowledge base. Lam (2006) proposed a fuzzy ontology map by extending the crisp ontology with the fuzzy theory and graph theory. Also, some efforts were made to construct fuzzy ontologies from various sources such as fuzzy narrower terms, fuzzy relations, and fuzzy database models as mentioned in Zhang et al. (2013). Moreover, fuzzy ontologies are applied to various application domains such as the Semantic Web (one of the most important applications), information retrieval, data mining, and context management (Bobillo 2008).

With the wide investigation and development of fuzzy XML, a significant interest developed regarding the problem of describing fuzzy XML with expressive knowledge representation techniques in recent years, so that some fuzzy XML issues such as reasoning and querying may be handled intelligently. After representing fuzzy XML in fuzzy *DLs*/ontologies, it is now possible to use their potential reasoning mechanism to reason on the fuzzy XML models. For instance, the reasoning tasks of fuzzy XML models (e.g., whether a fuzzy XML document conforms to a given document structure or whether a document structure is contained in another document structure) may be detected automatically through the reasoning mechanism of fuzzy *DLs*/ontologies instead of checking them by hand. Also, the correspondences may contribute to some tasks related to fuzzy XML models, such as integrating fuzzy XML documents and performing several optimization steps in answering queries over document bases. Therefore, in this chapter, we focus on how to reason on fuzzy XML with the knowledge representation formalisms fuzzy *DLs* and fuzzy ontologies, we transform fuzzy XML into fuzzy DL and fuzzy ontology, respectively, and then reason on fuzzy XML with the transformed formalisms.

7.2 Fuzzy XML Reasoning with Fuzzy Description Logic

For reasoning on fuzzy XML with fuzzy DL, a precondition here is that we first should represent fuzzy XML with fuzzy DL. Therefore, we need a fuzzy DL which has the good expressive power and reasoning ability for representing and reasoning on fuzzy XML. Note that the existing fuzzy *DLs* (e.g., *f-ALC*, *f-SHIN*, and *f-ALCIQ* as mentioned in Sect. 7.1), which have the limited expressive power, cannot account for the essential features of fuzzy XML models. In Zhang et al. (2011b), a fuzzy DL called f-ALCQ$_{wf\text{-}reg}$ is developed for the specific purposes of representing and reasoning on fuzzy XML models, where fuzzy XML models are translated into f-ALCQ$_{wf\text{-}reg}$ knowledge bases. Based on the translated f-ALCQ$_{wf\text{-}reg}$ knowledge bases, how to reason on fuzzy XML models (e.g., conformance, inclusion, equivalence, and disjointness) through the reasoning mechanism of f-ALCQ$_{wf\text{-}reg}$ is investigated. Also it is briefly discussed that how to support query processing over a document base more efficiently based on the translation and reasoning results above. In the following, we first introduce the fuzzy DL f-ALCQ$_{wf\text{-}reg}$, which is then used to represent and reason on fuzzy XML.

7.2.1 *Fuzzy Description Logic f-ALCQ$_{wf\text{-}reg}$*

In the following, the syntax, semantics, knowledge base, and reasoning algorithm are introduced for f-ALCQ$_{wf\text{-}reg}$.

7.2.1.1 The Syntax and Semantics of f-ALCQ$_{\mathbf{wf\text{-}reg}}$

Before introducing the fuzzy DL f-ALCQ$_{wf\text{-}reg}$, we first briefly recall and introduce some basic notions commonly occurring in DLs. In DLs, elementary descriptions are atomic concepts and atomic roles (also called concept names and role names), and complex descriptions can be built from them inductively with concept constructors and role constructors. The basic fuzzy DL is called *f-ALC* (Straccia 1998), and many fuzzy DLs are developed based on the further extension of *f-ALC*.

Let N_I, N_C, and N_R be three disjoint sets: N_I is a set of individual names, N_C is a set of fuzzy concept names, and N_R is a set of fuzzy role names. *f-ALC*-concepts are defined as follows (where $A \in N_C$, $R \in N_R$):

$C, D \rightarrow \top \mid$	(universal concept)
$\bot \mid$	(bottom concept)
$A \mid$	(atomic concept)
$\neg C \mid$	(concept negation)
$C \sqcap D \mid$	(intersection)
$C \sqcup D \mid$	(union)
$\exists R.C \mid$	(full existential quantification)
$\forall R.C$	(value restriction)

The semantics of *f-ALC* is defined by a fuzzy interpretation $FI = <\Delta^{FI}, \bullet^{FI}>$ where Δ^{FI} is a nonempty set and $\bullet^{FI}$ is a function which maps every $d \in N_I$ to an element $d^{FI} \in \Delta^{FI}$, maps every $A \in N_C$ into a function A^{FI}: $\Delta^{FI} \rightarrow [0, 1]$, and maps every $R \in N_R$ into a function R^{FI}: $\Delta^{FI} \times \Delta^{FI} \rightarrow [0, 1]$. Furthermore, for any *f-ALC*-concepts C and D, $R \in N_R$, and $d, d' \in \Delta^{FI}$, we have:

$$\top^{FI}(d) = 1$$
$$\bot^{FI}(d) = 0$$
$$(\neg C)^{FI}(d) = 1 - C^{FI}(d)$$
$$(C \sqcap D)^{FI}(d) = \min\{C^{FI}(d), D^{FI}(d)\}$$
$$(C \sqcup D)^{FI}(d) = \max\{C^{FI}(d), D^{FI}(d)\}$$
$$(\forall R.C)^{FI}(d) = \inf_{d' \in \Delta^{FI}}\{\max\{1 - R^{FI}(d, d'), C^{FI}(d')\}\}$$
$$(\exists R.C)^{FI}(d) = \sup_{d' \in \Delta^{FI}}\{\min\{R^{FI}(d, d'), C^{FI}(d')\}\}.$$

With the introduction of fuzzy sets into the classical *ALC*, the form of the knowledge base is changed accordingly. An *f-ALC* knowledge base *K* is composed of a TBox *T* and an ABox *A*:

1. A TBox *T* is a finite set of terminology axioms of the form $C \sqsubseteq D$ or $C = D$. An interpretation *FI* satisfies $C \sqsubseteq D$ iff for any $d \in \Delta^{FI}$, $C^{FI}(d) \leq D^{FI}(d)$, and similarly for $C = D$. *FI* is a model of TBox *T* iff *FI* satisfies all axioms in *T*.
2. An ABox *A* is a finite set of assertions of the form $<\alpha \bowtie n>$, Here $\bowtie \in \{>, \geq, <, \leq\}$, $n \in [0, 1]$, α is either of the form d: C or (d_1, d_2): R. Especially, in order to give a uniform format of the ABox, we define: when $n = 1$, the form $<\alpha \geq 1>$ is equivalent to $<\alpha = 1>$. Concretely speaking, $<d : C \geq 1>$ means that d is determinately an individual of C; $<(d_1, d_2) : R \geq 1>$ means that (d_1, d_2) determinately has the relationship R. An interpretation *FI* satisfies $<d: C \bowtie n>$ iff $C^{FI}(d^{FI}) \bowtie n$ and satisfies $<(d_1, d_2): R \bowtie n>$ iff $R^{FI}(d_1^{FI}, d_2^{FI}) \bowtie n$ iff $R^{FI}(d_1^{FI}, d_2^{FI}) \bowtie n$. *FI* is a model of ABox *A* iff *FI* satisfies all assertions in *A*.

A fuzzy interpretation *FI* satisfies an *f-ALC* knowledge base *K* if it satisfies all axioms and assertions in *K*.

Table 7.1 The syntax of fuzzy DL f-ALCQ$_{wf\text{-}reg}$

Construct	Syntax	Symbol
Top concept	$\top$	ALC
Bottom concept	$\bot$	
Atomic concept	A	
Concept negation	$\neg C$	
Conjunction	C $\sqcap$ D	
Value restriction	$\forall$R.C	
Exists restriction	$\exists$R.C	
Disjunction	C $\sqcup$ D	
Qualified number restrictions	$\geq$n R.C	Q
	$\leq$n R.C	
Well-founded	*Wf* (*R*)	*wf*
Atomic role	*P*	*reg*
Role union	$R_1 \sqcup R_2$	
Role composition	$R_1 \circ R_2$	
Refl. trans. closure	R*	
Transitive closure	R^+	
Identity	id (C)	

Based on the fuzzy DL *f-ALC*, the syntax of f-ALCQ$_{wf\text{-}reg}$ is shown in Table 7.1, the concepts and roles in f-ALCQ$_{wf\text{-}reg}$ are formed according to the syntax in Table 7.1. In Table 7.1:

- *A* denotes an atomic concept, *P* an atomic role, *C*, *D* arbitrary concept expressions, *R* an arbitrary role expression, *n* a non-negative integer.
- The concept construct *wf* (*R*) denotes the object that is the initial point of a sequence of roles.
- The role identity *id* (*C*) allows one to build a role that connects each instance of *C* to itself.

The semantics of f-ALCQ$_{wf\text{-}reg}$ are provided by a fuzzy interpretation $FI = (\Delta^{FI}, \bullet^{FI})$, where Δ^{FI} is a set of objects and $\bullet^{FI}$ is a fuzzy interpretation function, which maps:

- An individual d to an element $d^{FI} \in \Delta^{FI}$,
- A concept C to a membership function C^{FI}: $\Delta^{FI} \rightarrow [0, 1]$,
- A role R to a membership function R^{FI}: $\Delta^{FI} \times \Delta^{FI} \rightarrow [0, 1]$,
- The *semantics* is depicted as follows (c_i, d_i, x, y, $z \in \Delta^{FI}$):

$$\top^{FI}(d) = 1$$

$$\bot^{FI}(d) = 0$$

$$(\neg C)^{FI}(d) = 1 - C^{FI}(d)$$

$$(C \sqcap D)^{FI}(d) = \min\{C^{FI}(d), D^{FI}(d)\}$$

$$(C \sqcup D)^{FI}(d) = \max\{C^{FI}(d), D^{FI}(d)\}$$

$$(\forall R.C)^{FI}(d) = \inf_{d' \in \Delta^{FI}} \left\{ \max\left\{ 1 - R^{FI}\left(d, d'\right), C^{FI}\left(d'\right) \right\} \right\}$$

$$(\exists R.C)^{FI}(d) = \sup_{d' \in \Delta^{FI}} \left\{ \min\left\{ R^{FI}\left(d, d'\right), C^{FI}\left(d'\right) \right\} \right\}$$

$$(\geq nR.C)^{FI}(d) = \sup_{c_1, \ldots, c_n \in \Delta^{FI}} \wedge_{i=1}^{n} \{ \min\{ R^{FI}(d, c_i), C^{FI}(c_i) \} \}$$

$$(\leq nR.C)^{FI}(d) = \inf_{c_1, \ldots, c_{n+1} \in \Delta^{FI}} \vee_{i=1}^{n+1} \{ \max\{ 1 - R^{FI}(d, c_i), C^{FI}(c_i) \} \}$$

$$wf(R)^{FI}(d_0) = \begin{cases} 1 & if \ \forall d_1, d_2, \ldots, (ad \text{ infinitum}) \ \exists \ i \geq 0{:}R^{FI}(d_i, d_{i+1}) < \beta \\ 0 & else \end{cases}$$

$$id(C)^{FI}(d, d) = C^{FI}(d)$$

$$(R^*)^{FI} = \left(R^{FI}\right)^* = \cup_{n \geq 0} \left(R^{FI}\right)^n$$

$$(R^+)^{FI} = (R \circ R^*)^{FI}$$

$$(R_1 \sqcup R_2)^{FI}(x, y) = \max\left\{ R_1^{FI}(x, y), R_2^{FI}(x, y) \right\}$$

$$(R_1 \circ R_2)^{FI}(x, z) = \sup_{y \in Y} \left\{ \min\left\{ R_1^{FI}(x, y), R_2^{FI}(y, z) \right\} \right\}$$

The *concept construct wf* (*R*) is interpreted as those objects that are the initial point of finite *R*-chains, and membership degree $\beta \in (0, 1]$, which is a threshold value given according to the need of practical applications, is used to impose finiteness and acyclicity of all chains of objects. For example, if $\beta = 0.01$, then $R^{FI}(d_i, d_{i+1}) < 0.01$ denotes that there is no individual d_{i+1} which is a filler of the role *R* for d_i, i.e., d_i is the last point of the *R*-chain. Note that one can eliminate "+" from a role expression by replacing any occurrence of R^+ with $R \circ R^*$.

7.2.1.2 The Knowledge Base of f-ALCQ$_{wf\text{-}reg}$

An *f*-ALCQ$_{wf\text{-}reg}$ Knowledge Base (*f*-ALCQ$_{wf-reg}\mathcal{KB}$) consists of the *TBox* and the *ABox*. The Tbox introduces terminology, while the Abox contains assertions about the individuals.

A TBox T is a finite set of fuzzy axioms: inclusions $C \sqsubseteq D$ or equalities $C \equiv D$, where *C* and *D* are concepts. The semantics are interpreted as follows:

- $C \sqsubseteq D$, iff $\forall d \in \Delta^{FI}, C^{FI}(d^{FI}) \leq D^{FI}(d^{FI})$;
- $C \equiv D$, iff $\forall d \in \Delta^{FI}, C^{FI}(d^{FI}) = D^{FI}(d^{FI})$.

A fuzzy interpretation *FI* satisfies an *f*-ALCQ$_{wf\text{-}reg}$ TBox T iff it satisfies all fuzzy concept axioms in T; in this case, we say that *FI* is a model of T. Moreover, a Tbox T is called *simple* (Straccia 2001; Stoilos et al. 2005) if it neither includes cyclic nor general concept inclusions, i.e., axioms are the form $A \sqsubseteq C$ or $A \equiv C$, where A is a concept name that is never defined by itself either directly or indirectly, and A appears at most once at the left hand side.

An ABox A is a finite set of fuzzy assertions: concept assertions $<C(d) \bowtie n>$, and role assertions $<R(d_1, d_2) \bowtie n>$, where $\bowtie \in \{\geq, >, \leq, <\}$, $n \in [0, 1]$. The semantics are interpreted as follows:

- A fuzzy interpretation *FI* satisfies $<C(d) \bowtie n>$, iff $C^{FI}(d^{FI}) \bowtie n$;
- A fuzzy interpretation *FI* satisfies $<R(d_1, d_2) \bowtie n>$, iff $R^{FI}(d_1^{FI}, d_2^{FI}) \bowtie n$.

A fuzzy interpretation *FI* satisfies an *f*-ALCQ$_{wf\text{-}reg}$ ABox A iff it satisfies all fuzzy assertions in A; in this case, we say that *FI* is a model of A.

An *f*-ALCQ$_{wf\text{-}reg}$ knowledge base Σ is a pair $<$T, A$>$, a fuzzy interpretation *FI* satisfies Σ if *FI* satisfies all axioms and assertions in Σ; in this case, *FI* is called a *model* of Σ.

7.2.1.3 Reasoning of f-ALCQ$_{\mathbf{wf\text{-}reg}}$

In the following, we first introduce the basic reasoning problems of *f*-ALCQ$_{wf\text{-}reg}$. Then, we give a reasoning algorithm.

As in the case of fuzzy *DLs*, the basic tasks we consider when reasoning over an *f*-ALCQ$_{wf\text{-}reg}$ knowledge base $\Sigma = <\mathcal{T}, \mathcal{A}>$ are as follows:

- *Concept satisfiability*: A fuzzy concept *C* is *satisfiable* w.r.t. a TBox $\mathcal{T}$ iff there exists some model *FI* of $\mathcal{T}$ for which there is some $d \in \Delta^{FI}$ such that $C^{FI}(d^{FI}) = n$, $n \in (0, 1]$.
- *Concept subsumption*: A fuzzy concept *C* is *subsumed* by a concept *D* (written as $C \sqsubseteq D$) w.r.t. a TBox $\mathcal{T}$ if for every model *FI* of $\mathcal{T}$ it holds that, $\forall d \in \Delta^{FI}$. $C^{FI}(d^{FI}) \leq D^{FI}(d^{FI})$.
- *ABox consistency*: An *f*-ALCQ$_{wf\text{-}reg}$ ABox $\mathcal{A}$ is *consistent* w.r.t. a TBox $\mathcal{T}$ if there is a model *FI* of $\mathcal{T}$ which is also a model of $\mathcal{A}$.
- *Entailment*: An *f*-ALCQ$_{wf\text{-}reg}$ knowledge base Σ *entails* a fuzzy concept axiom or a fuzzy assertion Ψ, written $\Sigma \models \Psi$, iff all models of Σ satisfy Ψ.

In particular, in the presence of only *simple TBoxes*, all the problems above can be reduced to *ABox consistency* w.r.t. an empty TBox (Straccia 2001; Stoilos et al. 2005).

On this basis, the following briefly gives a reasoning algorithm for checking the *f*-ALCQ$_{wf\text{-}reg}$ ABox consistency. We first introduce a concept, i.e., *conjugated pairs of fuzzy assertions*.

Let ψ be a fuzzy assertion, with ψ^c indicates a *conjugate* of ψ (if there exists one). The conjugated pairs contain 4 forms:

$$\{<\alpha \geq n>, <\alpha < m>, n \geq m\},$$
$$\{<\alpha > n>, <\alpha < m>, n \geq m\},$$
$$\{<\alpha \geq n>, <\alpha \leq m>, n > m\},$$
$$\{<\alpha > n>, <\alpha \leq m>, n \geq m\},$$

where α denotes the assertion form $C(d)$ or $R(d_1, d_2)$, $n, m \in [0, 1]$.

Algorithm The ABox consistency reasoning algorithm.

Input: An f-ALCQ$_{wf\text{-}reg}$ ABox $\mathcal{A}$

Output: Boolean (true/false).

Suppose: (1) Assuming all concepts C occurring in $\mathcal{A}$ to be in *negation normal form* (NNF) (Straccia 2001; Stoilos et al. 2005), i.e., negations occur in front of concept names only; (2) The symbols $\bowtie$ denotes $\geq, >, \leq, <$; $\rhd$ denotes $\geq, >$; $\lhd$ denotes $\leq, <$; furthermore, we use symbols $\bowtie_r$, $\rhd_r$, and $\lhd_r$ to denote their reflections, for example the reflection of $\geq$ is $\leq$ and that of $>$ is $<$.

1. Applying the following transformation rules to the $\mathcal{A}$ until no more rules apply:

- $\neg\bowtie$: $(\neg C)(a) \bowtie n \in \mathcal{A}, C(a) \bowtie_r (1-n) \notin \mathcal{A} \to \mathcal{A} = \mathcal{A} \cup \{C(a) \bowtie_r (1-n)\}$;
- $\sqcap\rhd$: $(C \sqcap D)(a) \rhd n \in \mathcal{A}, C(a) \rhd n \notin \mathcal{A}, D(a) \rhd n \notin \mathcal{A} \to \mathcal{A} = \mathcal{A} \cup \{C(a) \rhd n, D(a) \rhd n\}$;
- $\sqcap\lhd$: $(C \sqcap D)(a) \lhd n \in \mathcal{A}, C(a) \lhd n \notin \mathcal{A}, D(a) \lhd n \notin \mathcal{A} \to \mathcal{A} = \mathcal{A} \cup \{C'\}, C' \in \{C(a) \lhd n, D(a) \lhd n\}$;
- $\sqcup\rhd$: $(C \sqcup D)(a) \rhd n \in \mathcal{A}, C(a) \rhd n \notin \mathcal{A}, D(a) \rhd n \notin \mathcal{A} \to \mathcal{A} = \mathcal{A} \cup \{C'\}, C' \in \{C(a) \rhd n, D(a) \rhd n\}$;
- $\sqcup\lhd$: $(C \sqcup D)(a) \lhd n \in \mathcal{A}, C(a) \lhd n \notin \mathcal{A}, D(a) \lhd n \notin \mathcal{A} \to \mathcal{A} = \mathcal{A} \cup \{C(a) \lhd n, D(a) \lhd n\}$;
- $\forall\rhd$: $\forall R.C(a) \rhd n \in \mathcal{A}, \exists b. \psi^c \in \mathcal{A}$ and $C(b) \rhd n \notin \mathcal{A}$, where $\psi = R(a,b) \rhd_r (1-n) \to \mathcal{A} = \mathcal{A} \cup \{C(b) \rhd n\}$;
- $\forall\lhd$: $\forall R.C(a) \lhd n \in \mathcal{A}, \neg\exists b. R(a,b) \lhd_r (1-n) \in \mathcal{A}$ and $C(b) \lhd n \in \mathcal{A} \to \mathcal{A} = \mathcal{A} \cup \{R(a,b) \lhd_r (1-n), C(b) \lhd n\}$;
- $\exists\rhd$: $\exists R.C(a) \rhd n \in \mathcal{A}, \neg\exists b. R(a,b) \rhd n \in \mathcal{A}$ and $C(b) \rhd n \in \mathcal{A} \to \mathcal{A} = \mathcal{A} \cup \{R(a,b) \rhd n, C(b) \rhd n\}$;
- $\exists\lhd$: $\exists R.C(a) \lhd n \in \mathcal{A}, \exists b. \psi^c \in A$ and $C(b) \lhd n \notin \mathcal{A}$, where $\psi = R(a,b) \lhd n \to \mathcal{A} = \mathcal{A} \cup \{C(b) \lhd n\}$;
- wf_1 : $wf(R)(b_0) = 1 \in \mathcal{A}, \neg\exists b_1, \cdots, b_n, b_{n+1} (n \geq 1).R(b_0, b_1) \rhd \beta \in \mathcal{A}, \ldots, R(b_{n-1}, b_n) \rhd \beta \in \mathcal{A}, R(b_n, b_{n+1}) < \beta \in \mathcal{A} \to \mathcal{A} = \mathcal{A} \cup \{R(b_0, b_1) \rhd \beta, \ldots, R(b_{n-1}, b_n) \rhd \beta\}$;
- $\leq\rhd$: $\leq kR.C(a) \rhd n \in \mathcal{A}, \exists b_1, \ldots, b_{k+1} (b_1 \neq b_2 \ldots \neq b_{k+1}).\psi_1^c \in \mathcal{A}, \ldots, \psi_{k+1}^c \in \mathcal{A}, C(b_1) \rhd n \notin \mathcal{A}, \ldots, C(b_{k+1}) \rhd n \notin \mathcal{A}$, and $R(a, b_1) \rhd_r (1-n) \notin \mathcal{A}, \ldots, R(a, b_{k+1}) \rhd_r (1-n) \notin \mathcal{A}$, where $\psi_1 = R(a, b_1) \rhd n, \ldots, \psi_{k+1} = R(a, b_{k+1}) \rhd n \to \mathcal{A} = \mathcal{A} \cup \{R(a, b_1) \rhd_r (1-n), \ldots, R(a, b_{k+1}) \rhd_r (1-n), C(b_1) \rhd n, \ldots, C(b_{k+1}) n\}$;
- $\geq\lhd$: $\geq kR.C(a) \lhd n \in \mathcal{A} \to$ apply $\leq\rhd$ rule to $\leq (k-1)R.C(a) \lhd_r (1-n)$;

- $\geq \triangleright$: $\geq kR.C(a) \triangleright n \in \mathcal{A}, \neg\exists b_1, \ldots, b_k(b_1 \neq b_2 \ldots \neq b_k).R(a, b_i) \triangleright n \in \mathcal{A}$ and $C(b_i) \triangleright n \in \mathcal{A} \rightarrow \mathcal{A} = \mathcal{A} \cup \{R(a, b_i) \triangleright n, \ldots, C(b_i) \triangleright n\}, i \in \{1 \ldots k\}$;
- $\leq \triangleleft$: $\leq kR.C(a) \triangleleft n \in \mathcal{A} \rightarrow$ apply $\geq \triangleleft$ rule to $\geq (k+1)R.C(a) \triangleleft_r (1-n)$;
- $id(C) \bowtie$: $id(C)(a, a) \bowtie n \in \mathcal{A}$ and $C(a) \bowtie n \notin \mathcal{A} \rightarrow \mathcal{A} = \mathcal{A} \cup \{C(a) \bowtie n\}$;
- $R^{*/+}$: $R^*(a, b) \bowtie n \in \mathcal{A}$, since $(R^*)^{FI} = (R^{FI})^* = \bigcup_{n \geq 0}(R^{FI})^n$ and $(R^+)^{FI} = (R \circ R^*)^{FI} \rightarrow R^{*/+}$ can be done by the following R_o and $R_\sqcup$ rules;
- $R_o \triangleright$: $(R_1 \circ R_2)(a, c) \triangleright n \in \mathcal{A}, \neg\exists b.R_1(a, b) \triangleright n \in \mathcal{A}$ and $R_2(b, c) \triangleright n \in \mathcal{A} \rightarrow \mathcal{A} = \mathcal{A} \cup \{R_1(a, b) \triangleright n, R_2(b, c) \triangleright n\}$;
- $R_o \triangleleft$: $(R_1 \circ R_2)(a, c) \triangleleft n \in \mathcal{A}, \neg\exists b.R_1(a, b) \triangleleft n \in \mathcal{A}$ and $R_2(b, c) \triangleleft n \in \mathcal{A} \rightarrow \mathcal{A} = \mathcal{A} \cup \{R^\ominus\}$, where $R^\ominus \in \{R_1(a, b) \triangleleft n, R_2(b, c) \triangleleft n\}$;
- $R_\sqcup \triangleright$: $(R_1 \sqcup R_2)(a, b) \triangleright n \in \mathcal{A}, R_1(a, b) \triangleright n \notin \mathcal{A}, R_2(a, b) \triangleright n \notin \mathcal{A} \rightarrow \mathcal{A} = \mathcal{A} \cup \{R^\ominus\}$, where $R^\ominus \in \{R_1(a, b) \triangleright n, R_2(a, b) \triangleright n\}$;
- $R_\sqcup \triangleleft$: $(R_1 \sqcup R_2)(a, b) \triangleleft n \in \mathcal{A}, R_1(a, b) \triangleleft n \notin \mathcal{A}, R_2(a, b) \triangleleft n \notin \mathcal{A} \rightarrow \mathcal{A} = \mathcal{A} \cup \{R_1(a, b) \triangleleft n, R_2(a, b) \triangleleft n\}$.

2. *Checking clash*: by applying the above rules until no more rules apply, we can obtain finite sets of fuzzy assertions $S = \{\mathcal{A}_1, \ldots, \mathcal{A}_k\}$ instead of single $\mathcal{A}$. If there is one of them such that it does not contain *clash*, then f-ALCQ$_{wf\text{-}reg}$ ABox $\mathcal{A}$ is *consistent*, and $\mathcal{A}$ is *inconsistent* otherwise. Here, a set contains a *clash* if at least one occurs in the following 10 situations, where C is a fuzzy concept, $d \in \Delta^{FI}$:
   - The four conjugated pairs above;
   - Bottom concept $\perp$ or top concept $\top$:

     $\{\perp(d) \geq n,\ n > 0\}$,
     $\{\perp(d) > n,\ n > 0\}$,
     $\{\top(d) \leq n,\ n < 1\}$,
     $\{\top(d) < n,\ n < 1\}$.
   - $\{C(d) > 1\}, \{C(d) < 0\}$.

3. End.

Notice that, the algorithm is given based on the existing reasoning algorithms for *f-ALC* (Straccia 2001), *f-SHIN* (Stoilos et al. 2005), *f-ALCIQ* (Stoilos et al. 2008), and *ALCIQ$_{reg}$* (Giacomo and Lenzerini 1994), while some new rules are presented in order to handle all f-ALCQ$_{wf\text{-}reg}$ concept and role constructors. Here, the algorithm is a decision procedure for the consistency of f-ALCQ$_{wf\text{-}reg}$ ABox $\mathcal{A}$ which can be drawn following the similar procedures given in Straccia (2001), Stoilos et al. (2005, 2008), Giacomo and Lenzerini (1994).

Based on the fuzzy DL f-ALCQ$_{wf\text{-}reg}$, next we introduce how to represent and reason on fuzzy XML models with f-ALCQ$_{wf\text{-}reg}$.

7.2.2 *Representing Fuzzy XML with Fuzzy Description Logic f-ALCQ$_{wf\text{-}reg}$*

The precondition for reasoning on fuzzy XML is that we first should represent fuzzy XML with fuzzy DL. In this section, we introduce how to represent fuzzy XML with the fuzzy *DL f*-ALCQ$_{wf\text{-}reg}$. We give some rules to transform the fuzzy XML into fuzzy *DL f*-ALCQ$_{wf\text{-}reg}$ knowledge bases, and implement a prototype transformation tool.

7.2.2.1 Transforming Fuzzy XML into f-ALCQ$_{wf\text{-}reg}$ Knowledge Bases

The following Definition 7.1 gives a formal approach for transforming a fuzzy XML model into an *f*-ALCQ$_{wf\text{-}reg}$ *KB*. We now show that the structural aspects of fuzzy XML document can be captured in *f*-ALCQ$_{wf\text{-}reg}$ by defining a translation function φ from the corresponding fuzzy DTD to *f*-ALCQ$_{wf\text{-}reg}$ knowledge base, then establishing a correspondence between fuzzy document instances and models of the derived knowledge base. Starting with the construction of *atomic fuzzy concepts* and *atomic fuzzy roles*, the approach induces *a set of fuzzy axioms* from the fuzzy XML model.

Definition 7.1 Let (*D*, *d*) be a fuzzy XML model, where $D = (\mathbf{P}, r)$ is a fuzzy DTD as mentioned in Definition 3.1 in Chap. 3, **P** is a set of *element type definitions*, *r* is the *root element type*, and *d* is the *fuzzy XML document instance* conforming to the fuzzy DTD *D*. The *f*-ALCQ$_{wf\text{-}reg}$ *KB* $\varphi(D) = <FA, FT>$ can be defined by function φ as follows:

1. The *FA* of φ (*D*), which is a set of *atomic fuzzy concepts* and *atomic fuzzy roles*, contains the following elements:

- For each element type definition $E \rightarrow \alpha \in \mathbf{P}$, creating one unique atomic fuzzy concept E_D. The same element type *E* in different fuzzy DTDs D_1 and D_2 corresponds to the distinct atomic fuzzy concepts E_{D1} and E_{D2}.
- For each element type $E \in \mathbf{E}$, creating two atomic fuzzy concepts Start*E* and End*E*, which represent respectively the start tag and end tag of *E*. Specially, the pairs of atomic fuzzy concepts (Start*Val*, End*Val*), (Star*Poss*, End*Poss*), (Start*Dist*, End*Dist*), (Start*Type*, End*Type*), are the particular atomic fuzzy concepts in *FA*, which used for the representation of fuzzy data in the fuzzy XML document and are independent from the specific fuzzy DTD.
- For each terminal *T* (such as #PCDATA), creating one atomic fuzzy concept *T*.
- Creating two extra atomic fuzzy concepts **Tag** and **Terminal**, to distinguish between *tags* and *terminals*.
- Creating the extra atomic fuzzy roles **f** and **r**.

2. The *FT* of φ (*D*), which is a set of *fuzzy axioms*, contains $FT_{\mathbf{d}}$ and $FT_{\mathbf{D}}$ in the following:

- $FT_{\mathbf{d}}$: *A set of fuzzy XML document instance axioms, used to restrict the properties of the fuzzy XML document d.*
  $FT_{\mathbf{d}}$ is used to capture the general structural properties of fuzzy document instances, i.e., enforce that every model of $FT_{\mathbf{d}}$ represents a fuzzy document instance *d* by means of a tree. We can translate a fuzzy document instance into a binary tree (i.e., each node of the tree has at most two children nodes) following the steps (*a–c*).

(a) The *root* of the tree, which uniquely specifies a fuzzy XML document instance, corresponds to the root element of the fuzzy XML document instance *d*.
(b) The role **f** represents the start tag of an element and **r** represents the other components of the element.
(c) In detail, for a fuzzy XML document instance *d* with *n* components, i.e., <*E*> <E_1>···</E_1>···<E_n>···</E_n> </*E*>, the *start tag* is represented by the **f**-filler of the root, the *first component* by the (**r**o**f**)-filler, …, the *last component* by the ($\mathbf{r}^n$o**f**)-filler, and the *end tag* by the $\mathbf{r}^{n+1}$-filler, which can be illustrated as follows:

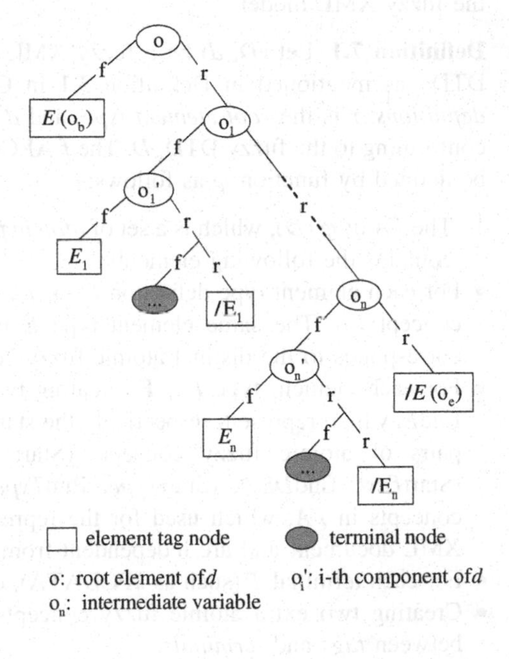

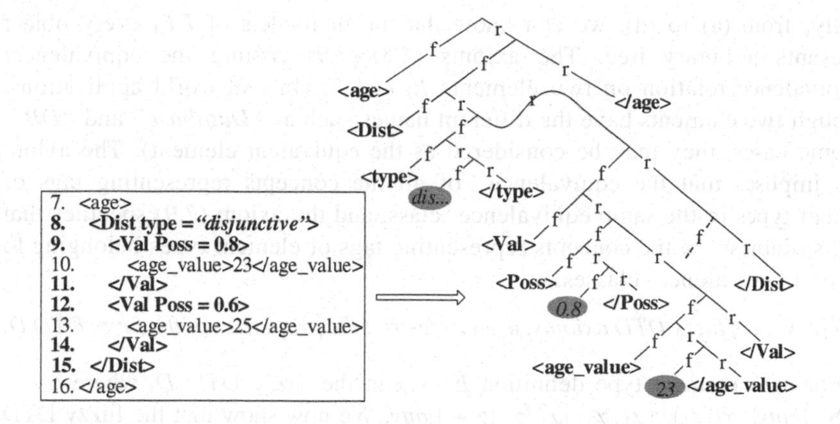
7. <age>
8. <Dist type = "disjunctive">
9. <Val Poss = 0.8>
10. <age_value>23</age_value>
11. </Val>
12. <Val Poss = 0.6>
13. <age_value>25</age_value>
14. </Val>
15. </Dist>
16. </age>
<age>
</age>
<Dist>
<type>
dis...
</type>
<Val>
<Poss>
0.8
</Poss>
</Dist>
</Val>
<age_value>
23
</age_value>
f
r

(d) Based on the tree representation of a fuzzy XML document, i.e., (a–c), the set of fuzzy XML document instance axioms $FT_{\mathbf{d}}$ can be defined as follows:

$$\top \equiv (\leq 1\mathbf{f}.\top) \sqcap (\leq 1\mathbf{r}.\top) \sqcap wf(\mathbf{f} \sqcup \mathbf{r}) \tag{7.1}$$

$$\mathbf{Tag} \sqsubseteq \forall (\mathbf{f} \sqcup \mathbf{r}).\bot \tag{7.2}$$

$$\text{Start}E \sqsubseteq \mathbf{Tag} \tag{7.3}$$

$$\text{End}E \sqsubseteq \mathbf{Tag} \tag{7.4}$$

$$\mathbf{Terminal} \sqsubseteq \forall (\mathbf{f} \sqcup \mathbf{r}).\bot \sqcap \neg\, \mathbf{Tag} \tag{7.5}$$

$$T \sqsubseteq \mathbf{Terminal} \tag{7.6}$$

$$T_1 \sqsubseteq \neg T_2 \tag{7.7}$$

$$\begin{cases} \text{Start}E_i \equiv \text{Start}E_j & \text{where the tag names of elements} \\ \text{End}E_i \equiv \text{End}E_j & E_i \text{ and } E_j \text{ are equivalent} \end{cases} \tag{7.8}$$

$$\begin{cases} \text{Start}E_i \subseteq \neg\text{Start}E_j & \text{where the tag names of elements} \\ \text{End}E_i \subseteq \neg\text{End}E_j & E_i \text{ and } E_j \text{ are not equivalent} \end{cases} \tag{7.9}$$

Here, a fuzzy document instance is by definition finite, and hence has a finite nesting of components. While the *well-founded* construct $wf(R)$ is interpreted as those objects that are initial point of only finite R-chains, therefore, we can impose finiteness and acyclicity of all chains of objects connected by $\mathbf{f} \sqcup \mathbf{r}$. Additionally, T of axiom (7.6) denotes a terminal, the axiom (7.1) ensures us that in every model of $FT_{\mathbf{d}}$, every object is the root a tree in which every node has at most one **f** successor and one **r** successor, the axioms (7.2) and (7.5) show that **Tag** and **Terminal** are disjoint, and the instances of **Tag** and **Terminal** are leaves of the tree, the axioms (7.6) and (7.7) impose that two different terminals have disjoint instances, the axioms (7.3) and (7.4) specify Start$E$ and End$E$ are subsets of **Tag**.

Finally, from (a) to (d), we can know that in all models of $FT_{\mathbf{d}}$ every object represents a binary tree. The axioms (7.8)–(7.9) capture the equivalence/inequivalence relation on two elements E_i and E_j (In real-world applications, although two elements have the different names such as "*Database*" and "*DB*", in some cases, they may be considered as the equivalent element). The axiom (7.8) imposes that the equivalence of all the concepts representing tags of element types in the same equivalence class, and the axiom (7.9) specifies that the disjointness of the concepts representing tags of element types belonging to different equivalence classes.

- $FT_{\mathbf{D}}$: *A set of fuzzy DTD axioms, used to restrict the properties of the fuzzy DTD D.*

(a) For each element type definition $E \rightarrow \alpha$ in the fuzzy DTD D, where $\alpha ::= \mathbf{S}|empty|\,(\alpha_1|\alpha_2)\,|\,(\alpha_1, \alpha_2)\,|\alpha?\,|\alpha^*\,|\alpha+\,|\,any$. We now show that the fuzzy DTD can be captured in *f*-ALCQ$_{wf\text{-}reg}$ by defining a translation φ from element type definition $E \rightarrow \alpha \in \mathbf{P}$ in fuzzy DTD $D = (\mathbf{P}, r)$ to $FT_{\mathbf{D}}$, for each element type definition $E \rightarrow \alpha \in \mathbf{P}$, $FT_{\mathbf{D}}$ contains the fuzzy axiom:

$$E_{\mathrm{D}} \equiv \exists \mathbf{f}.\ \mathrm{Start}E \sqcap \exists\,(\mathbf{r} \circ \varphi(\alpha)\,).\ \mathrm{End}E$$

where φ (α) reflects the structure restrictions about $E \rightarrow \alpha \in \mathbf{P}$ and are defined as follows:

$$\begin{aligned}
\varphi(\mathbf{S}) &= \mathrm{id}\ (\exists \mathbf{f}.\beta(D, \mathbf{S})\,)\mathrm{o}\ \mathbf{r} \\
\varphi(\alpha_1|\alpha_2) &= \varphi(\alpha_1) \sqcup \varphi(\alpha_2) \\
\varphi(\alpha_1, \alpha_2) &= \varphi(\alpha_1)\mathrm{o}\ \varphi(\alpha_2) \\
\varphi(empty) &= \mathrm{id}\ (\top) \\
\varphi(\alpha^*) &= \varphi(\alpha) \sqcup \mathrm{id}\ (\top) \\
\varphi(\alpha^*) &= \varphi(\alpha)^* \\
\varphi(\alpha^+) &= \varphi(\alpha)^+.
\end{aligned}$$

Here, the mapping function β $(D, \mathbf{S})$ is used to establish the relationships between elements in fuzzy DTD such that $(\mathbf{S} = \mathbf{T} \cup \mathbf{E})$: if $\mathbf{S} = E$ for an element type $E \in \mathbf{E}$, $\beta(D, \mathbf{S}) = E_{\mathrm{D}}$; if $\mathbf{S} = T$ for a terminal $T \in \mathbf{T}$, β $(D, \mathbf{S}) = T$.

(b) The following gives the *f*-ALCQ$_{wf\text{-}reg}$ knowledge base derived from a fragment of fuzzy DTD (which is basically from the fuzzy DTD D_1 mentioned in Fig. 3.2). Note that the general parts and some axioms are omitted.

Universities$_{\mathrm{D}}$ ≡ ∃**f**.Start*Universities* ⊓ ∃(**r** o (id(∃**f**.*University*$_{\mathrm{D}}$) o **r**)*). End *Universities*;

University$_{\mathrm{D}}$ ≡ ∃**f**.Start*University* ⊓ ∃(**r** o id(∃**f**.*UName*$_{\mathrm{D}}$) o **r** o id(∃**f**.*Val*$_{\mathrm{D1}}$) o **r** o (id(∃**f**.*Val*$_{\mathrm{D1}}$) o **r**)*). End*University*;

UName$_{\mathrm{D}}$ ≡ ∃**f**.Start*UName* ⊓ ∃(**r** o id(∃**f**.#PCDATA) o **r**). End*UName*;

Val$_{\mathrm{D1}}$ ≡ ∃**f**.Start*Val* ⊓ ∃(**r** o id(∃**f**.*Poss*$_{\mathrm{D}}$) o **r** o (id(∃**f**.*Student*$_{\mathrm{D}}$) o **r**)*). End*Val*;

$Student_D \equiv \exists$**f**.Start*Student* $\sqcap \exists$(**r** o ((id($\exists$**f**.$Sname_D$) o **r**) $\sqcup$ id($\top$)) o ((id ($\exists$**f**.Age_D) o **r**) $\sqcup$ id($\top$)) o ((id($\exists$**f**.$Email_D$) o **r**) $\sqcup$ id($\top$))). End*Student*;
$Sname_D \equiv \exists$**f**.Start*Sname* $\sqcap \exists$(**r** o id($\exists$**f**.#PCDATA) o **r**). End*Sname*;
$Dist_D \equiv \exists$**f**.Start*Dist* $\sqcap \exists$(**r** o id($\exists$**f**.$Type_D$) o **r** o id($\exists$**f**.Val_{D2}) o **r** o (id ($\exists$**f**.Val_{D2}) o **r**)*). End*Dist*;
⋮
$Email\_value_D \equiv \exists$**f**.Start*Email_value* $\sqcap \exists$(**r** o id($\exists$**f**.#PCDATA) o **r**). End*Email_value*.

Based on the Definition 7.1, we realize the formal translation from fuzzy DTD to *f*-$ALCQ_{wf\text{-}reg}$ knowledge base. Below we prove the correctness of the translation above, which can be sanctioned by establishing the mappings between the fuzzy XML document and the model of *f*-$ALCQ_{wf\text{-}reg}$ *KB* (Theorem 7.1).

Theorem 7.1 *For every fuzzy XML model (D, d), where D* = (**P**, *r) is a fuzzy DTD, and d is a fuzzy XML document instance conforming to D. The* φ *(D) is the corresponding f-*$ALCQ_{wf\text{-}reg}$ *KB derived by Definition 7.1. There exist mappings* $\boldsymbol{\mu}$ *from the fuzzy XML document instance to the model of* φ *(D), and* $\boldsymbol{\lambda}$ *from the model of* φ *(D) to the fuzzy XML document instance, such that*:

- For each fuzzy XML document instance *d*, there is $\boldsymbol{\mu}$ (*d*) which is a model of φ (*D*), and for each $d \in d_{\mathbf{T,E}}$ iff $o \in \beta$ $(D, \mathbf{S})^{\boldsymbol{\mu}(d)}$, where *o* is the root of $\boldsymbol{\mu}$ (*d*);
- For each model *FI* of φ (*D*), and $o \in \Delta^{FI}$, there is $\boldsymbol{\lambda}$ (*o*) which is a fuzzy XML document instance for *D*, and for each $o \in \beta$ $(D, \mathbf{S})^{FI}$ iff $\boldsymbol{\lambda}(o) \in d_{\mathbf{T,E}}$.

Proof The following briefly gives the proof of Theorem 7.1. Given two alphabets **T** (the basic types such as #PCDATA and CDATA) and **E** (element types), all fuzzy XML document instances $d_{\mathbf{T,E}}$ built over **T** and **E** can be defined inductively as follows: (1) If *d* is a terminal in **T**, then $d_i \in d_{\mathbf{T,E}}$; (2) If *d* is a sequence of the form $<E> d_1, \ldots, d_k </E>$, where $E \in \mathbf{E}$ is an element type and $d_1, \ldots, d_k \in d_{\mathbf{T,E}}$, then $d \in d_{\mathbf{T,E}}$. Firstly, for part 1, let *d* be a fuzzy XML document, then a model of $\varphi(D)\boldsymbol{\mu}(d) = \left(\Delta^{\mu(d)}, \bullet^{\mu(d)}\right)$ can be defined inductively:

(a) If *d* is a terminal $T \in \mathbf{T}$, then $\Delta^{\boldsymbol{\mu}(d)} = T^{\boldsymbol{\mu}(d)} = \mathbf{Terminal}^{\boldsymbol{\mu}\ (d)}$;
(b) If *d* is a sequence of the form $<E> d_1, \ldots, d_n </E>$, then:

$$\Delta^{\mu(d)} = \{o, o_b, o_1, \ldots, o_n, o_e\} \cup \bigcup_{1 \le i \le n} \Delta^{\mu(d_i)}$$

$$StartE^{\mu(d)} = \{o_b\} \cup \bigcup_{1 \le i \le n} StartE^{\mu(d_i)}$$

$$EndE^{\mu(d)} = \{o_e\} \cup \bigcup_{1 \le i \le n} EndE^{\mu(d_i)}$$

$$\mathrm{Tag}^{\mu(d)} = \{o_b, o_e\} \cup \bigcup_{1 \le i \le n} \mathrm{Tag}^{\mu(d_i)}$$

$$f^{\mu(d)} = \left\{(o, o_b), \left(o_1, o_1'\right), \ldots, \left(o_n, o_n'\right)\right\} \cup \bigcup_{1 \le i} f^{\mu(d_i)}$$

$$r^{\mu(d)} = \{(o, o_1), (o_1, o_2), \ldots, (o_{n-1}, o_n), (o_n, o_e)\} \cup \bigcup_{1 \le i \le n} r^{\mu(d_i)}$$

where *o* is the root of *d*, o_b and o_e denote the start and end tags of element *E*, o_i denotes the *i*-th component of *d*, and o_i' is the root of d_i, $i \in \{1, \ldots, n\}$;

and for part 2, which can be proved similarly for the first part above. Let $FI = (\Delta^{FI}, \bullet^{FI})$ be a model of φ (*D*) and $o \in \Delta^{FI}$, a fuzzy document instance λ (*o*) for *D* can be defined:

(a) If $o \in T^{FI}$ for some terminal $T \in \mathbf{T}$, then $\lambda(o) = T$.
(b) If for some $E \in \mathbf{E}$, there are some integer $n \geq 0$ and objects o_b, o_i, o_i', and o_e, such that $o_b \in \text{Start}E^{FI}, o_e \in \text{End}E^{FI}$, (o, o_b), $(o_1, o_1'), \ldots, (o_n, o_n') \in \mathbf{f}^{FI}$, and $(o, o_1), (o_1, o_2), \ldots, (o_{n-1}, o_n), (o_n, o_e) \in \mathbf{r}^{FI}$, then $\lambda(o) = \langle E \rangle\, \lambda(o_1'), \ldots, \lambda(o_n') \langle /E \rangle$. □

7.2.2.2 Prototype Translation Tool

Following the proposed approach above, we developed a prototype translation tool called *FXML2DL*, which can automatically translate fuzzy XML models into *f*-ALCQ$_{wf\text{-}reg}$ knowledge bases. In the following, we briefly introduce the design and implementation of *FXML2DL*.

The implementation of *FXML2DL* is based on Java 2 JDK 1.5 platform. *FXML2DL* includes three main modules: *parsing module*, *translation module*, and *output module*. The parsing module parses the fuzzy XML document files and fuzzy DTDs and stores the parsed results as Java ArrayList classes; The translation module uses Java class methods to translate fuzzy XML documents and fuzzy DTDs into *f*-ALCQ$_{wf\text{-}reg}$ knowledge bases based on the proposed approach; The output module produces the resulting *f*-ALCQ$_{wf\text{-}reg}$ knowledge bases which are saved as the text files and displayed on the tool screen. The overall architecture of *FXML2DL* is shown in Fig. 7.1. It should be noted that *FXML2DL* cannot automatically extract the fuzzy DTD according to the fuzzy XML document if there is not a fuzzy DTD in hand.

Here we give an example to show that the proposed approach is feasible and the tool is efficient. We carried out transformation experiments of some fuzzy XML

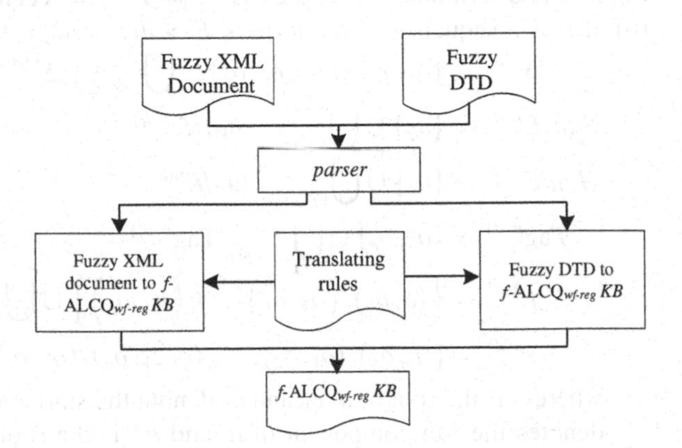

Fig. 7.1 The architecture of *FXML2DL*

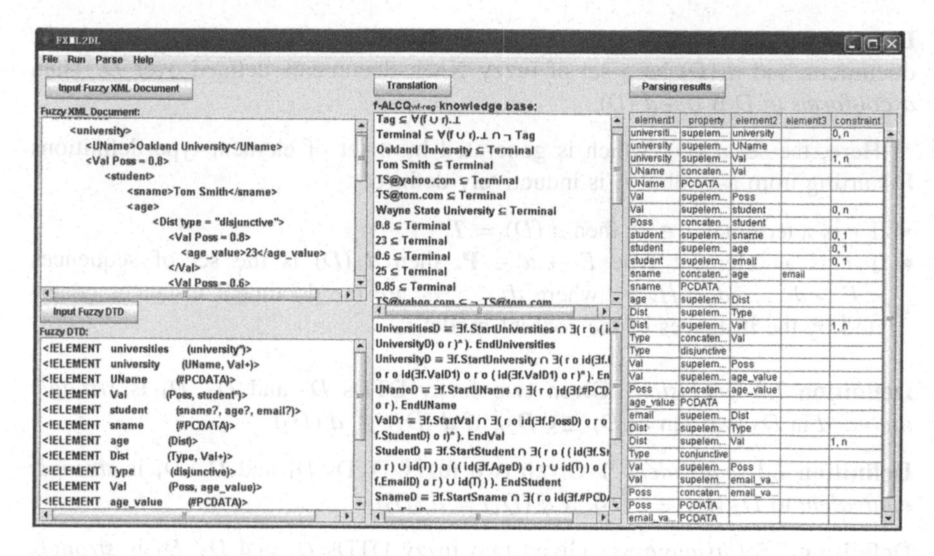

Fig. 7.2 The screen snapshot of *FXML2DL*

models using the implemented tool *FXML2DL*, with a PC (CPU P4/3.0 GHz, RAM 3.0 GB and Windows XP system). Figure 7.2 shows the screen snapshot of *FXML2DL*, which displays the transformation from a fuzzy XML document and fuzzy DTD mentioned in Chap. 3 to the corresponding f-ALCQ$_{wf\text{-}reg}$ knowledge base.

7.2.3 *Reasoning on Fuzzy XML with Fuzzy Description Logic f-ALCQ$_{wf\text{-}reg}$*

Based on the transformed f-ALCQ$_{wf\text{-}reg}$ *KB*, in this section we further introduce how to reason on fuzzy XML models with the transformed f-ALCQ$_{wf\text{-}reg}$ *KB*. The basic reasoning tasks of fuzzy XML models are briefly introduced, and then the approach for reasoning on fuzzy XML models with f-ALCQ$_{wf\text{-}reg}$ is developed.

7.2.3.1 Reasoning Problems of Fuzzy XML Models

The familiar reasoning problems considered in fuzzy XML models include conformance, inclusion, disjointness, and equivalence. In the following, we give the formal definitions of these reasoning problems based on the reasoning tasks on classical XML (Buneman et al. 2003; Baader et al. 2003; Wu et al. 2008; Cautis et al. 2007; Libkin and Sirangelo 2008; Toman and Weddell 2005; Calvanese et al. 1999).

Definition 7.2 (*conformance*) Let $D = (\mathbf{P}, r)$ be a fuzzy DTD, d be a fuzzy XML document, and d (D) be a set of fuzzy XML documents defined over D. Then *d conforms to D* if $d \in d\ (D)$.

Here, the set d (D), which is generated by a set of element type definitions **P** starting from a symbol r, is inductively defined:

- If r is a terminal $T \in \mathbf{T}$, then $d\ (D) = T$.
- If r is an element type $E \to \alpha \in \mathbf{P}$, then d (D) is the set of sequences $<E> d_1, \ldots, d_k </E>$, where $d_1, \ldots, d_k$ are the document instances generated by the instances of the content model α.

Definition 7.3 (*inclusion*) Given two fuzzy DTDs D_1 and D_2. D_1 is *strongly included* in D_2 (written as $D_1 \subseteq_S D_2$), if $d\ (D_1) \subseteq d\ (D_2)$.

Definition 7.4 (*equivalence*) Given two fuzzy DTDs D_1 and D_2. D_1 is *strongly equivalent* to D_2 ($D_1 \equiv_S D_2$), if $d\ (D_1) = d\ (D_2)$.

Definition 7.5 (*disjointness*) Given two fuzzy DTDs D_1 and D_2. D_1 is *strongly disjoint* from D_2 ($D_1 \otimes_S D_2$), if $d\ (D_1) \cap d\ (D_2) = \emptyset$.

Notice that, for determining *strong* inclusion (equivalence, disjointness) between fuzzy DTDs, i.e., determining whether two fuzzy DTDs define the inclusive (same, disjoint) sets of fuzzy XML document instances, the names of start/end tags of elements in document instances play a key role. For example, given two fuzzy DTDs (part of elements only):

$D_{\text{University}}$ <!ELEMENT *University* (Uname, Position, Val+, …)>
D_{College} <!ELEMENT *College* (Cname, Position, Val+, …)>,
based on the Definitions 7.2–7.5, we know that $D_{\text{University}}$ is *strongly disjoint* from D_{College}.

In real-world applications, it appears that the restriction imposed by *strong* is rather *limited*. In some cases, although the tags of elements in two fuzzy DTDs are *different*, we know that the structures of two fuzzy DTDs are *related*. For example, if we consider that the tag names *University* and *Uname* are equivalent to *College* and *Cname*, respectively, then $D_{\text{University}}$ is *equivalent* to D_{College}.

On this basis, by considering the *equivalent/inequivalent relation* of tag names, i.e., certain tag names as equal, and others as different, one can determine the inclusion (equivalence, etc.) relationships between fuzzy DTDs well, which are called *ε-inclusion* (*ε-equivalence*, etc.). The *strong inclusion* (*equivalence*, etc.) are just special cases of *ε-inclusion* (*ε-equivalence*, etc.).

Therefore, in the following, we consider the reasoning problems of *ε-inclusion* (*ε-equivalence*, etc.) only, and their formal definitions are given as follows.

Definition 7.6 (*ε-conformance*) Let $D = (\mathbf{P}, r)$ be a fuzzy XML DTD, d be a fuzzy XML document instance, and $d_\varepsilon(D)$ be a set of fuzzy ε-document instances defined over D. Then d ε-conforms to D if $d \in d_\varepsilon(D)$.

Here, the set $d_\varepsilon(D)$ is defined following the similar procedures in Definition 7.2:

- If r is a terminal $T \in \mathbf{T}$, then $d_\varepsilon(D) = T$.
- If r is an element type $E \rightarrow \alpha \in \mathbf{P}$, then $d_\varepsilon(D)$ is the set of sequences $<E'> d_1, \ldots, d_k </E'>$, where $E' \in [E]_\varepsilon$ ($[E]_\varepsilon$ denotes the set of equivalence classes of E), and $d_1, \ldots, d_k$ are the document instances generated by the instances of the content model α.

Definition 7.7 (*ε-inclusion*) Given two fuzzy DTDs D_1 and D_2. D_1 is *ε-included* in D_2 (written as $D_1 \subseteq_\varepsilon D_2$), if $d_\varepsilon(D_1) \subseteq d_\varepsilon(D_2)$.

Definition 7.8 (*ε-equivalence*) Given two fuzzy DTDs D_1 and D_2. D_1 is *ε-equivalent* to D_2 ($D_1 \equiv_\varepsilon D_2$), if $d_\varepsilon(D_1) = d_\varepsilon(D_2)$.

Definition 7.9 (*ε-disjointness*) Given two fuzzy DTDs D_1 and D_2. D_1 is *ε-disjoint* from D_2 ($D_1 \otimes_\varepsilon D_2$), if $d_\varepsilon(D_1) \cap d_\varepsilon(D_2) = \varnothing$.

7.2.3.2 Reasoning on Fuzzy XML Model with f-ALCQ$_{wf\text{-}reg}$

The following theorems allow us to reduce reasoning on fuzzy XML models to reasoning on *f*-ALCQ$_{wf\text{-}reg}$ *KB*, so that the above reasoning tasks of fuzzy XML models may be reasoned by means of the reasoning mechanism of *f*-ALCQ$_{wf\text{-}reg}$.

Theorem 7.2 (ε-conformance) *Let D* = (**P**, *r*) *be a fuzzy DTD, d be a fuzzy XML document instance, and φ (D) be the f*-ALCQ$_{wf\text{-}reg}$ *KB derived by Definition* 7.1. *Then checking whether d ε-conforms to D can be reduced to model checking in φ (D), i.e., checking whether μ (d) in Theorem 7.1 is a model of φ (D).*

It is clear that Theorem 7.2 is an immediate consequence of Definition 7.1 and Theorem 7.1. The proof of Theorem 7.2 is omitted here, which can be drawn easily by Theorem 7.1.

In order to determine *ε-inclusion* (*ε-equivalence*, and *ε-disjointness*) between fuzzy DTDs, in the following, we first need to extend the notion of *f*-ALCQ$_{wf\text{-}reg}$ *KB* φ (D) in such a way that φ (D) is derived from a set of fuzzy DTDs, rather than a single fuzzy DTD (see Definition 7.10).

Definition 7.10 Given a set $D = \{D_1, \ldots, D_n\}$ of fuzzy DTDs, the *f*-ALCQ$_{wf\text{-}reg}$ *KB* φ (D) is defined as $\varphi(D) = \varphi(D_1) \cup \cdots \cup \varphi(D_n)$ (FT_D part only), which can be derived by Definition 7.1.

Theorem 7.3 (ε-inclusion) *Given a set* $D = \{D_1, D_2\}$ *of fuzzy DTDs, where* $D_1 = (\mathbf{P}_1, r')$ *and* $D_2 = (\mathbf{P}_2, r'')$, r' *and* r'' *are the root element types,* $\varphi(D) = \varphi(D_1) \cup \varphi(D_2)$ *is the f*-ALCQ$_{wf\text{-}reg}$ *KB derived by Definition* 7.1, r'_{D1} *and* r''_{D2} *are two atomic fuzzy concepts in* φ (D). *Then* $D_1 \subseteq_\varepsilon D_2$ iff $\varphi(D) \models r'_{D1} \sqsubseteq r''_{D2}$.

Proof "⇒": If $\varphi(D) \nvDash \mathrm{r}'_{D1} \sqsubseteq r''_{D2}$, then there is $o \in \Delta^{FI}$ such that $o \in \mathrm{r}'^{FI}_{D1}$ and $o \notin \mathrm{r}''^{FI}_{D2}$, where $FI = (\Delta^{FI}, \bullet^{FI})$ is a model of φ (D). By the part 2 of Theorem 7.1, $\boldsymbol{\lambda}$ (o) is a fuzzy XML document instance, and $\boldsymbol{\lambda}(o) \in d_{\varepsilon}(D_1)$ and $\boldsymbol{\lambda}(o) \notin d_{\varepsilon}(D_2)$, i.e., $D_1 \nsubseteq_{\varepsilon} D_2$. Contradiction, so φ $(D) \vDash \mathrm{r}'_{D1} \sqsubseteq r''_{D2}$.

"⇐": If $D_1 \nsubseteq_{\varepsilon} D_2$, then there is a fuzzy XML document instance d with $d \in d_{\varepsilon}(D_1)$ and $d \notin d_{\varepsilon}(D_2)$. By the part 1 of Theorem 7.1, $\boldsymbol{\mu}$ (d) is a model of φ (D), and $\mathrm{r}'_{D1} \in \beta(D, \mathbf{S})^{\mu(d)}$ and $r''_{D2} \notin \beta(D, \mathbf{S})^{\mu(d)}$, i.e., φ $(D) \nvDash \mathrm{r}'_{D1} \sqsubseteq r''_{D2}$. Contradiction, so $D_1 \subseteq_{\varepsilon} D_2$. □

Theorem 7.4 (ε-equivalence) *Given a set* $D = \{D_1, D_2\}$ *of fuzzy DTDs, where* $D_1 = (\mathbf{P}_1, r')$ *and* $D_2 = (\mathbf{P}_2, r'')$, r' *and* r'' *are the root element types.* φ $(D) = \varphi$ $(D_1) \cup \varphi$ (D_2) *is the f-ALCQ*$_{wf\text{-}reg}$ *KB derived by Definition 7.1,* r'_{D1} *and* r''_{D2} *are two atomic fuzzy concepts in* φ (D)*. Then* $D_1 \equiv_{\varepsilon} D_2$ iff $\varphi(D) \vDash \mathrm{r}'_{D1} \equiv r''_{D2}$.

Proof "⇐": If $D_1 \not\equiv_{\varepsilon} D_2$, then there is a fuzzy XML document instance d with $d \in d_{\varepsilon}(D_1)$ and $d \notin d_{\varepsilon}(D_2)$ or $d \notin d_{\varepsilon}(D_1)$and $d \in d_{\varepsilon}(D_2)$. By the part 1 of Theorem 7.1, $\boldsymbol{\mu}$ (d) is a model of φ (D), and $\mathrm{r}'_{D1} \in \beta(D, \mathbf{S})^{\mu(d)}$ and $r''_{D2} \notin \beta(D, \mathbf{S})^{\mu(d)}$ or $\mathrm{r}'_{D1} \notin \beta(D, \mathbf{S})^{\mu(d)}$ and $r''_{D2} \in \beta(D, \mathbf{S})^{\mu(d)}$, i.e., φ $(D) \nvDash \mathrm{r}'_{D1} \equiv r''_{D2}$. Contradiction, so $D_1 \equiv_{\varepsilon} D_2$.

"⇒": If $\varphi(D) \nvDash \mathrm{r}'_{D1} \equiv r''_{D2}$, then there is $o \in \Delta^{FI}$ such that $o \in \mathrm{r}'^{\ FI}_{D1}$ and $o \notin \mathrm{r}''^{\ FI}_{D2}$ or $o \notin \mathrm{r}'^{\ FI}_{D1}$ and $o \in \mathrm{r}''^{\ FI}_{D2}$, where $FI = (\Delta^{FI}, \bullet^{FI})$ is a model of φ (D). By the part 2 of Theorem 7.1, $\boldsymbol{\lambda}$ (o) is a fuzzy XML document instance, and $\boldsymbol{\lambda}(o) \in d_{\varepsilon}(D_1)$ and $\boldsymbol{\lambda}(o) \notin d_{\varepsilon}(D_2)$ or $\boldsymbol{\lambda}(o) \notin d_{\varepsilon}(D_1)$ and $\boldsymbol{\lambda}(o) \in d_{\varepsilon}(D_2)$, i.e., $D_1 \not\equiv_{\varepsilon} D_2$. Contradiction, so φ $(D) \vDash \mathrm{r}'_{D1} \equiv r''_{D2}$. □

Theorem 7.5 (ε-disjointness). *Given a set* $D = \{D_1, D_2\}$ *of fuzzy DTDs, where* $D_1 = (\mathbf{P}_1, r')$ and $D_2 = (\mathbf{P}_2, r'')$, r'*and* r'' *are the root element types,* φ $(D) = \varphi$ $(D_1) \cup \varphi$ (D_2) *is the f-ALCQ*$_{wf\text{-}reg}$ *KB derived by Definition 7.1,* r'_{D1} *and* r''_{D2} *are two atomic fuzzy concepts in* φ (D)*. Then* $D_1 \otimes_{\varepsilon} D_2$iff φ $(D) \vDash \mathrm{r}'_{D1} \sqcap r''_{D2} \sqsubseteq \perp$.

Proof "⇒": If $\varphi(D) \nvDash \mathrm{r}'_{D1} \sqcap r''_{D2} \sqsubseteq \perp$, then there is $o \in \Delta^{FI}$ such that $o \in \mathrm{r}'^{\ FI}_{D1}$ and $o \in r''^{\ FI}_{D2}$, where $FI = (\Delta^{FI}, \bullet^{FI})$ is a model of φ (D). By the part 2 of Theorem 7.1, $\boldsymbol{\lambda}$ (o) is a fuzzy XML document instance, and $\boldsymbol{\lambda}$ $(o) \in d_{\varepsilon}(D_1)$ and $\boldsymbol{\lambda}$ $(o) \in d_{\varepsilon}(D_2)$, i.e., $d_{\varepsilon}(D_1)_{\varepsilon}(D_2) \neq \varnothing$. Contradiction, so φ $(D) \vDash \mathrm{r}'_{D1} \sqcap r''_{D2} \sqsubseteq \perp$.

"⇐": If D_1 is not *ε-disjoint* from D_2, i.e., $d_{\varepsilon}(D_1) \cap d_{\varepsilon}(D_2) \neq \varnothing$, then there is a fuzzy XML document instance d with $d \in d_{\varepsilon}(D_1)$ and $d \in d_{\varepsilon}(D_2)$. By the part 1 of Theorem 7.1, $\boldsymbol{\mu}$ (d) is a model of φ (D), and $\mathrm{r}'_{D1} \in \beta$ $(D, \mathbf{S})^{\mu(d)}$ and $r''_{D2} \in \beta$ $(D, \mathbf{S})^{\mu(d)}$, i.e., φ $(D) \nvDash \mathrm{r}'_{D1} \sqcap r''_{D2} \sqsubseteq \perp$. Contradiction, so $D_1 \otimes_{\varepsilon} D_2$. □

In order to illustrate that representation and reasoning of fuzzy XML models with the fuzzy *DL* may contribute to some tasks related to fuzzy XML models, for example, improving the precision and efficiency of query processing over document bases, based on the previous translation and reasoning results, the following briefly discusses how to perform several optimization steps in answering queries over a document base. The detailed introduction about the query of fuzzy XML documents is not included here.

A document base *DCB* is a pair $DCB = <D_s, d_\varepsilon(D_s)>$, where D_s is a set of fuzzy DTDs, and $d_\varepsilon(D_s)$ is a set of fuzzy ε-document instances defined over D_s. A query *Q* over a document base *DCB* is simply a fuzzy document type definition (i.e., a fuzzy DTD), used to retrieve all document instances that *ε-conform to Q*, i.e., $Q\ (DCB) = \{d \mid d \in d_\varepsilon(D_s) \wedge d \in d_\varepsilon(Q)\}$.

The following Rules 1–5 can perform several optimization steps in answering query *Q* over a document base *DCB*.

Firstly, based on our previous work, with the assumption that for each pair fuzzy DTDs $D_i, D_j \in D_s, i \neq j$, it is known that whether $D_i \subseteq_\varepsilon D_j$, $D_i \equiv_\varepsilon D_j$, or $D_i \otimes_\varepsilon D_j$; for each pair $d \in d_\varepsilon(D_s)$, $D_i \in D_s$, it is known that whether *d ε-conforms to* D_i (i.e., $d \in d_\varepsilon(D_i)$ or $d \notin d_\varepsilon(D_i)$).

Rule 1: If we know that there is a fuzzy DTD $D_1 \in D_s$ such that $D_1 \equiv_\varepsilon Q$ (where D_1 is selected from D_s such that there is no $D_2 \in D_s$ with $D_1 \subseteq_\varepsilon D_2$), then *Q(DCB)* is the set of the document instances that *ε-conform to* D_1;

Rule 2: If we know that there is a fuzzy DTD $D_1 \in D_s$ such that $D_1 \otimes_\varepsilon Q$ (where D_1 is selected ditto), then we can discard all fuzzy DTDs that are *ε-included in* D_1, and exclude from the answer all document instances that *ε-conform to* D_1;

Rule 3: If we know that there is a fuzzy DTD $D_1 \in D_s$ such that $D_1 \subseteq_\varepsilon Q$ (where D_1 is selected ditto), then each $d \in d_\varepsilon(D_1)$ takes part in the answer to the query. In addition, each D' (where $D' \in D_s$ and $D' \subseteq_\varepsilon D_1$) needs not to be considered anymore and is discarded;

Rule 4: If we know that there is a fuzzy DTD $D_1 \in D_s$ such that $Q \subseteq_\varepsilon D_1$ (where D_1 is selected ditto), that is, each $d \in d_\varepsilon(Q)$ also satisfies $d \in d_\varepsilon(D_1)$, then we can discard all document instances that *ε-conform to* D' (where $D' \in D_s$ and $D' \otimes_\varepsilon D_1$);

Rule 5: If there are no fuzzy DTD $D_1 \in D_s$ that satisfy the conditions in the Rules 1–4, then we can remove D_1 from D_s.

7.3 Fuzzy XML Reasoning with Fuzzy Ontology

Being similar to the fuzzy XML reasoning with fuzzy Description Logic, in order to reason on fuzzy XML with fuzzy ontology, a precondition here is also that we first can represent fuzzy XML with fuzzy ontology. In Zhang et al. (2013), fuzzy ontologies are constructed from fuzzy XML data resources, and then the constructed fuzzy ontologies are used to reason on fuzzy XML. In this section, we focus on fuzzy XML reasoning with fuzzy ontology.

7.3.1 Representing Fuzzy XML with Fuzzy Ontology

To represent fuzzy XML with fuzzy ontology, here we first briefly introduce some basic notions about fuzzy ontology. In a more general sense, the short answer for this question "what is a fuzzy ontology" is: a fuzzy ontology is a shared model of some domain which is often conceived as a hierarchical data structure containing all concepts, properties, individual, and their relationships in the domain, where these concepts, properties and so on may be defined imprecisely. Ontology can be defined by ontology representation languages such as RDFS, OIL, DAML+OIL, or OWL. Web Ontology language (OWL), a W3C recommendation, has been the Semantic Web de-facto standard language for representing ontologies (OWL 2004). OWL has three increasingly expressive sublanguages OWL Lite, OWL DL, and OWL Full. Description Logics are the main logical foundations of the ontology OWL. Although OWL is a quite expressive formalism it features limitations, mainly with what can be said about imprecise and uncertain information that is commonly found in real-world applications. Therefore, the fuzzy extension of OWL, i.e., fuzzy OWL, was developed. Table 7.2 shows the fuzzy OWL syntax and the corresponding fuzzy DL syntax. In Table 7.2, C denotes fuzzy class description (i.e., fuzzy DL concept); D denotes fuzzy data range (i.e., fuzzy DL concrete datatype); R denotes fuzzy ObjectProperty identifier (i.e., fuzzy DL abstract role); U denotes fuzzy DatatypeProperty identifier (i.e., fuzzy DL concrete role), d and o are abstract individuals; v is a concrete individual; n is a nonnegative integer, $\#\{\}$ denotes the base of a set $\{\}$, and $\bowtie \in \{\geq, >, \leq, <\}$. The semantics for fuzzy OWL is given based on the interpretation of the fuzzy DL *f-SHOIN(D)* (Straccia 2005). Briefly, the semantics is provided by a fuzzy interpretation $FI = (\Delta^{\mathrm{FI}}, \Delta_D, o^{\mathrm{FI}}, o^{\mathrm{D}})$, where Δ^{FI} is the abstract domain and Δ_D is the datatype domain (disjoint from Δ^{FI}), and $\bullet^{\mathrm{FI}}$ and $\bullet^{\mathrm{D}}$ are two fuzzy interpretation functions, which map:

- An abstract individual o to an element $o^{\mathrm{FI}} \in \Delta^{\mathrm{FI}}$,
- A concrete individual v to an element $v^{\mathrm{D}} \in \Delta_D$,
- A concept A to a membership degree function $A^{\mathrm{FI}} : \Delta^{\mathrm{FI}} \rightarrow [0, 1]$,
- An abstract role R to a membership degree function $R^{\mathrm{FI}} : \Delta^{\mathrm{FI}} \times \Delta^{\mathrm{FI}} \rightarrow [0, 1]$,
- A concrete datatype D to a membership degree function $D^{\mathrm{D}} : \Delta_D \rightarrow [0, 1]$,
- A concrete role U to a membership degree function $U^{\mathrm{FI}} : \Delta^{\mathrm{FI}} \times \Delta_D \rightarrow [0, 1]$.

Based on the fuzzy interpretation *FI*, the complete semantics of fuzzy OWL abstract syntax can be found in (Zhang et al. 2013).

A fuzzy ontology can be formulated in the fuzzy OWL language. Formally, a fuzzy ontology FO consists of the fuzzy ontology structure FO_{S} and the corresponding fuzzy ontology instance FO_{I}. FO_{S} is a set of *identifiers* and *fuzzy class/property axioms*, and FO_{I} is a set of *fuzzy individual axioms*. The following Definition 7.11 gives a formal definition of fuzzy ontology (Zhang et al. 2013).

Table 7.2 The fuzzy OWL syntax and its corresponding fuzzy Description Logic (*DL*) syntax

Fuzzy OWL syntax	The corresponding fuzzy DL Syntax
Fuzzy class descriptions (C)	
A, which is a URIref of a class	A
Owl:Thing	$\top$
Owl:Nothing	$\bot$
IntersectionOf (C_1, …, C_n)	$C_1 \sqcap \cdots \sqcap C_n$
UnionOf (C_1, …, C_n)	$C_1 \sqcup \cdots \sqcup C_n$
ComplementOf (C)	$\neg C$
Restriction (R allValuesFrom(C))	$\forall R.C$
Restriction (R minCardinality(n))	$\geq n\ R$
Restriction (R maxCardinality(n))	$\leq n\ R$
Restriction (R cardinality(n))	$= n\ R$
Restriction (U allValuesFrom(D))	$\forall U.D$
Fuzzy class axioms	
Class (A partial C_1, …, C_n)	$A \sqsubseteq C_1 \sqcap \cdots \sqcap C_n$
Class (A complete C_1, …, C_n)	$A = C_1 \sqcap \cdots \sqcap C_n$
SubClassOf (C_1 C_2)	$C_1 \sqsubseteq C_2$
EquivalentClasses (C_1, …, C_n)	$C_1 \equiv \cdots \equiv C_n$
DisjointClasses (C_1, …, C_n)	$C_i \neq C_j$
Fuzzy property axioms	
DatatypeProperty (U	
domain(C_1)…domain(C_m)	$\geq 1U \sqsubseteq C_i$
range(D_1)…range(D_k)	$\top \sqsubseteq \forall U.D_i$
[Functional])	$\top \sqsubseteq\ \leq 1U$
ObjectProperty (R	
domain(C_1)…domain(C_m)	$\geq 1R \sqsubseteq C_i$
range(C_1)… range(C_k)	$\top \sqsubseteq \forall R.C_i$
[Functional]	$\top \sqsubseteq\ \leq 1R$
[inverseOf (R_0)])	$R = (R_0)^-$
Fuzzy individual axioms	
Individual (o type(C_1) [$\bowtie m_i$]	$o : C_i \bowtie m_i$
value(R_1, o_1) [$\bowtie k_i$]…	(o, o_i): $R_i \bowtie k_i$
value(U_1, v_1) [$\bowtie l_i$]…)	(o, v_i): $U_i \bowtie l_i$
SameIndividual (o_1, …, o_n)	$o_1 = \cdots = o_n$
DifferentIndividuals (o_1, …, o_n)	$o_i \neq o_j$

Definition 7.11 (*fuzzy ontology*) A fuzzy ontology is a tuple $FO = (FO_S, FO_I) = (FID_O, FAxiom_O)$, where:

1. $FID_O = FCID_O \cup FIID_O \cup FDRID_O \cup FOPID_O \cup FDPID_O$ is a finite fuzzy OWL identifier set partitioned into:

   - A subset $FCID_O$ of fuzzy class identifiers, include user-defined identifiers plus two predefined fuzzy classes owl: Thing and owl: Nothing.
   - A subset $FIID_O$ of individual identifiers.
   - A subset $FDRID_O$ of fuzzy data range identifiers.

- A subset $FOPID_O$ of fuzzy object property identifiers; fuzzy object properties link individuals to individuals.
- A subset $FDPID_O$ of fuzzy datatype property identifiers; fuzzy datatype properties link individuals to data values.

2. $FAxiom_O$ is a finite fuzzy OWL axiom set partitioned into:
   - A subset of fuzzy class/property axioms, used to represent fuzzy ontology structure.
   - A subset of fuzzy individual axioms, used to represent fuzzy ontology instance.

In brief, a fuzzy ontology *FO* is a set of fuzzy axioms defined on the identifier, and we say that a fuzzy interpretation *FI* is a model of *FO* iff it satisfies all axioms in *FO*.

In order to represent fuzzy XML with fuzzy ontologies, two key steps need to be done: *first*, transforming the fuzzy DTD into a fuzzy ontology at structure level; *second*, on the basis of the first step, transforming the fuzzy XML document into the fuzzy ontology at instance level.

As we have known, the DTD and the ontology solve the different problems: the former provides means to express and constrain the syntax and structure of XML documents, the latter, in contrast, is intended for modeling the semantic relationships of a domain. However, there is an interesting overlap between the two, as both of them have an object-oriented foundation and are based on the notion of Frames. On this basis, the following introduces how to establish the correspondences between fuzzy DTDs and fuzzy ontologies at conceptual level.

In the following, we propose a formal approach (i.e. a set of rules) for transforming a fuzzy DTD into fuzzy ontology structure. Starting with the construction of *fuzzy identifiers* from the symbols in a fuzzy DTD (e.g., element names and attribute names), the rules further induce *a set of fuzzy axioms* (i.e., fuzzy class and property axioms) from the fuzzy DTD. The fuzzy axioms are obtained by mapping each element type definition $E \rightarrow (\alpha, A)$ of the fuzzy DTD in Chap. 3 into the corresponding fuzzy axioms, where $\alpha ::= (\mathbf{S} \mid \text{empty} \mid (\alpha_1|\alpha_2) \mid (\alpha_1,\alpha_2) \mid \alpha? \mid \alpha^* \mid \alpha+ \mid \text{any})$ and $A ::= (AN, AT, VT)$. Moreover, we give the corresponding example illustrations regarding some key transformation rules to explain the construction process well.

Given a fuzzy DTD $D = (P, r)$, i.e., a set of *element type definitions* $E \rightarrow (\alpha, A)$ in Definition 3.1 in Chap. 3. The fuzzy ontology structure $FO_S = \varphi(D) = (FID_O, FAxiom_O)$ can be derived by transformation function φ as shown in the following Rules 1–11:

Rule 1: Each element symbol *E* in element type definitions $E \rightarrow (\alpha, A)$ is mapped into a fuzzy class identifier $\varphi(E) \in FCID_O \in FID_O$.

For example, the element symbols such as *university* and *UName* in the fuzzy DTD D_1 of Fig. 3.2 are mapped into fuzzy class identifiers $\varphi(university) \in FCID_O$ and $\varphi(UName) \in FCID_O$, respectively.

Rule 2: Each type T, i.e., the atomic type for E (such as #PCDATA) or the attribute type for A (such as CDATA), is mapped into a fuzzy data range identifier φ (T) $\in FDRID_0 \in FID_0$.

Rule 3: For each attribute type definition $E \to A$, where A ::= (AN, AT, VT), AN is the attribute name, AT is the attribute type, and VT is a symbol "use", creating the following elements:

- creating a fuzzy datatype property identifier φ (EhasdpAN) $\in FDPID_0$, where φ (EhasdpAN) $= E +$ "hasdp" $+ AN$, which is used to denote the relationship between an element E and its associated attribute AN, and "+" denotes the concatenated operation of strings;
- creating the following fuzzy class and fuzzy datatype property axioms:

  Class (φ (E) partial restriction (φ (EhasdpAN) allValuesFrom (φ (AT)) minCardinality (use (VT)) maxCardinality (use (VT)))), where if VT = "#IMPLIED", then use (VT) = 0 or 1; if VT = "#REQUIRED", then use (VT) = 1; if VT = #Fixed "value" or "value", then φ (AT) is the "value" and use (VT) = ∞;
  DatatypeProperty (φ(EhasdpAN) domain (φ(E)) range (φ(AT))).

Rule 4: For each element type definition $E \to T$, creating the following elements:

- creating a fuzzy datatype property identifier φ(Ehasdp$T$) $\in FDPID_0$, which is used to denote the relationship between an element E and its type;
- creating the following fuzzy class and fuzzy datatype property axioms:

  Class (φ(E) partial restriction (φ(Ehasdp$T$) allValuesFrom (φ(T)) cardinality (1));
  DatatypeProperty (φ(Ehasdp$T$) domain (φ(E)) range (φ(T)) [Functional]).

For example, for the element type definition *UName* → (*#PCDATA*) in the fuzzy DTD D_1 of Fig. 3.2, first we create a fuzzy class identifier φ (*UName*) $\in FCID_0$ and a fuzzy data range identifier φ (*PCDATA*) $\in FDRID_0$ by using the rules 1 and 2, and then create a fuzzy datatype property identifier φ (*UName*hasdp*PCDATA*) $\in FDPID_0$ and the following axioms: Class (φ (*UName*) partial restriction (φ (*UName*hasdp*PCDATA*) allValuesFrom (φ (*PCDATA*)) cardinality (1))); DatatypeProperty (φ (*UName*hasdp*PCDATA*) domain (φ (*UName*)) range (φ (*PCDATA*)) [Functional]).

Rule 5: For each element type definition $E \to \alpha$, creating the following elements:

- creating a fuzzy object property identifier φ (Ehasopα) $\in FOPID_0$, where φ (Ehasopα) $= E +$ "hasop" $+ \alpha$, which is used to denote the relationship between an element E and its content model α, and "+" denotes the concatenated operation of strings;
- creating the following fuzzy class and fuzzy object property axioms:

Class (φ (E) partial restriction (φ (Ehasopα) allValuesFrom (φ (α)) Cardinality (1)));

ObjectProperty (φ (Ehasopα) domain (φ (E)) range (φ (α)) [Functional]).

Rule 6: For each element type definition $E \rightarrow \alpha?$, creating the following elements:

- creating a fuzzy object property identifier φ (Ehasopα) $\in$ $FOPID_0$, which is used to denote the relationship between an element E and its content model α;
- creating the following fuzzy class and fuzzy object property axioms:

Class (φ (E) partial restriction (φ (Ehasopα) allValuesFrom (φ (α)) maxCardinality (1)));

ObjectProperty (φ (Ehasopα) domain (φ (E)) range (φ (α))).

For example, for the element type definition *student* $\rightarrow$ *sname*?… in the fuzzy DTD D_1 of Fig. 3.2, first we create two fuzzy class identifiers φ(*student*), φ(*sname*) $\in$ $FCID_0$ by using the rule 1, and then create a fuzzy object property identifier φ (*student*hasop*sname*) $\in$ $FOPID_0$ and the following axioms: Class (φ (*student*) partial restriction (φ (*student*hasop*sname*) allValuesFrom (φ (*sname*)) maxCardinality (1)) …); ObjectProperty (φ (*student*hasop*sname*) domain (φ (*student*)) range (φ (*sname*))).

Rule 7: For each element type definition $E \rightarrow \alpha^*$, creating the following elements:

- creating a fuzzy object property identifier φ (Ehasopα) $\in$ $FOPID_0$, which is used to denote the relationship between an element E and its content model α;
- creating the following fuzzy class and fuzzy object property axioms:

Class (φ (E) partial restriction (φ (Ehasopα) allValuesFrom (φ (α))));

ObjectProperty (φ (Ehasopα) domain (φ (E)) range (φ (α))).

For example, for the element type definition *universities* $\rightarrow$ *university** in the fuzzy DTD D_1 of Fig. 3.2, first we create two fuzzy class identifiers φ (*universities*), φ (*university*) $\in$ $FCID_0$ by using the rule 1, and then create a fuzzy object property identifier φ (*universities*hasop*university*) $\in$ $FOPID_0$ and the following axioms: Class (φ (*universities*) partial restriction (φ (*universities*hasop*university*) allValuesFrom (φ (*university*)))); ObjectProperty (φ (*universities*hasop*uni versity*) domain (φ (*universities*)) range (φ (*university*))).

Rule 8: For each element type definition $E \rightarrow \alpha+$, creating the following elements:

- creating a fuzzy object property identifier φ (Ehasopα) $\in$ $FOPID_0$, which is used to denote the relationship between an element E and its content model α;
- creating the following fuzzy class and fuzzy object property axioms:

Class (φ (E) partial restriction (φ (Ehasopα) allValuesFrom (φ (α)) minCardinality (1)));

ObjectProperty (φ (Ehasopα) domain (φ (E)) range (φ (α))).

For example, for the element type definition *university* → (…, *Val*+) in the fuzzy DTD D_1 of Fig. 3.2, first we create two fuzzy class identifiers φ (*university*), φ (*Val*) ∈ $FCID_0$ by using the rule 1, and then create a fuzzy object property identifier φ (*university*hasop*Val*) ∈ $FOPID_0$ and the following axioms: Class (φ (*university*) partial restriction (φ (*university*hasop*Val*) allValuesFrom (φ (*Val*)) minCardinality (1))); ObjectProperty (φ (*university*hasop*Val*) domain (φ (*university*)) range (φ (*Val*))).

Rule 9: For each element type definition E → empty, creating the following fuzzy class axiom:

Class (φ (E) partial owl:Nothing).

Rule 10: For each element type definition $E \rightarrow (\alpha_1|\alpha_2)$, creating the following elements:

- creating two fuzzy object property identifiers φ (Ehasopα_1) ∈ $FOPID_0$ and φ (Ehasopα_2) ∈ $FOPID_0$, which are also used to denote the relationships between an element E and α_i;
- creating the following fuzzy class and fuzzy object property axioms:

  Class (φ (E) complete unionOf (intersectionOf (φ (α_1) complementOf (φ (α_2))) intersectionOf (complementOf (φ (α_1)) φ (α_2))));
  Class (φ (E) partial restriction (φ (Ehasopα_1) allValuesFrom (φ (α_1))) restriction (φ (Ehasopα_2) allValuesFrom (φ (α_2))));
  ObjectProperty (φ (Ehasopα_1) domain (φ (E)) range (φ (α_1)));
  ObjectProperty (φ (Ehasopα_2) domain (φ (E)) range (φ (α_2))).

For example, for an element type definition *lecture* → (*name* | *phone*) in a fuzzy DTD (which is not included in the fuzzy DTD D_1 of Fig. 3.2 and expresses that a *lecturer* element contains either a *name* element or a *phone* element), first we create three fuzzy class identifiers φ (*lecture*), φ (*name*), and φ (*phone*) ∈ $FCID_0$ by using the rule 1, and then create two fuzzy object property identifiers φ (*lecture*hasop*name*), φ (*lecture*hasop*phone*) ∈ $FOPID_0$ and the following axioms: Class (φ (*lecture*) complete unionOf (intersectionOf (φ (*name*) complementOf (φ (*phone*))) intersectionOf (complementOf (φ (*name*)) φ (*phone*)))); Class (φ (*lecture*) partial restriction (φ (*lecture*hasop*name*) allValuesFrom (φ (*name*))) restriction (φ (*lecture*hasop*phone*) allValuesFrom (φ (*phone*)))); ObjectProperty (φ (*lecture*hasop*name*) domain (φ (*lecture*)) range (φ (*name*))); ObjectProperty (φ (*lecture*hasop*phone*) domain (φ (*lecture*)) range (φ (*phone*))).

Rule 11: For each element type definition $E \rightarrow (\alpha_1,\alpha_2)$, creating the following elements:

- creating two fuzzy object property identifiers φ (Ehasopα_1) ∈ $FOPID_0$ and φ (Ehasopα_2) ∈ $FOPID_0$, which are used to denote the relationships between an element E and α_i;
- creating the following fuzzy class and fuzzy object property axioms:

Class (φ (E) complete intersectionOf (φ (α_1) φ (α_2)));
Class (φ (E) partial restriction (φ (Ehasopα_1) allValuesFrom (φ (α_1))) restriction (φ (Ehasopα_2) allValuesFrom (φ (α_2))));
ObjectProperty (φ (Ehasopα_1) domain (φ (E)) range (φ (α_1)));
ObjectProperty (φ (Ehasopα_2) domain (φ (E)) range (φ (α_2))).

For example, for the element type definition *student* → (*sname*?, *age*?, …) in the fuzzy DTD D_1 of Fig. 3.2, first we create three fuzzy class identifiers φ (*student*), φ (*sname*), and φ (*age*) $\in FCID_0$ by using the rule 1, and then create two fuzzy object property identifiers φ (*student*hasop*sname*), φ (*student*hasop*age*) $\in FOPID_0$ and the following axioms by jointly using the rules 6 and 11: Class (φ (*student*) complete intersectionOf (φ (*sname*) φ (*age*) …)); Class (φ (*student*) partial restriction (φ (*student*hasop*sname*) allValuesFrom (φ (*sname*)) maxCardinality (1)) restriction (φ (*student*hasop*age*) allValuesFrom (φ (*age*)) maxCardinality (1)) …); ObjectProperty (φ (*student*hasop*sname*) domain (φ (*student*)) range (φ(*sname*))); ObjectProperty (φ (*student*hasop*age*) domain (φ (*student*)) range (φ (*age*))).

Based on the transformation at the structure level above, the following further transforms fuzzy XML models into fuzzy ontologies at instance level.

Given a fuzzy XML document $d = (N, <, \lambda, \eta, r)$ in Definition 3.2 in Chap. 3 conforming to a fuzzy DTD D, the corresponding *fuzzy ontology instance* $FO_I = \varphi(d) = (FID_0, FAxiom_0)$ can be derived as the following Rules 12–14:

Rule 12: For each node $v \in N$ in the fuzzy XML document tree, if $\lambda(v) \in \mathbf{E}$, then creating a fuzzy individual identifier φ (v) $\in FIID_0 \in FID_0$ and an individual axiom: Individual (φ (v) type(φ ($\lambda(v)$)) [$\bowtie m_i$]).

Here: (1) φ ($\lambda(v)$) is a fuzzy class identifier $FCID_0$ created by the rule 1; (2) φ (v) = "*id*" + *six-digit number*, such as *id123456*, where "+" denotes the concatenated operation of strings, and the *six-digit number* is generated randomly according to the level of the node v in the fuzzy XML document tree and is unique in fuzzy ontology instances; (3) The above axiom aims at creating a unique individual for each fuzzy class in constructed fuzzy ontology structure, and the default value of membership degree m_i is 1.0. In the following, the part [$\bowtie$ 1.0] in a fuzzy individual axiom is omitted for the sake of similarity.

For example, for the nodes such as *university*, *UName* $\in N$ in the fuzzy XML document, we create two fuzzy individual identifiers *id592914*, *id593011* $\in FIID_0$ and two individual axioms: Individual (*id592914* Type (φ (*university*))); Individual (*id593011* Type (φ (*UName*))). Here, φ (*university*), φ (*UName*) are two created fuzzy class identifiers $FCID_0$.

Rule 13: For each pair nodes v_i, $v_j \in N$ in the fuzzy XML document tree, if $\lambda(v_i)$, $\lambda(v_j) \in \mathbf{E}$, and $v_i < v_j$, i.e., v_i is the parent node of v_j, then creating the axioms: Individual (φ (v_j) type (φ ($\lambda(v_j)$))) and Individual (φ (v_i) type (φ ($\lambda(v_i)$)) value (φ ($v_i$hasopv_j), φ (v_j))). Furthermore:

- if $\lambda(v_j) \in \mathbf{E}$ is a leaf element node, then adding the axiom: Individual (φ (v_j) type (φ ($\lambda(v_j)$)) value (φ ($v_j$hasdp*PCDATA*), $\eta(v_i, v_j)$)));
- If $\lambda(v_j) = Val_j$, $\lambda(v_i) \neq Dist_i$, and there is a node $v_h \in N$ such that $v_h < v_i$, then adding the axiom: Individual ($\varphi(v_h)$ type($\varphi(\lambda(v_h))$) value($\varphi(v_h$hasop$v_i)$, $\varphi(v_i)$) [$\bowtie m_j$]), where $m_j = \eta(v_j$, @Poss$_j)$;
- If $\lambda(v_j) = Val_j$, $\lambda(v_i) = Dist_i$, we have: (1) if there is a node $v_k \in N$ such that $v_j < v_k$ and $\lambda(v_k) \in \mathbf{E}$, then adding the axiom: Individual (φ (v_j) type (φ ($\lambda(v_j)$)) value (φ ($v_j$hasopv_k), φ (v_k)) [$\bowtie m_j$]), where $m_j = \eta(v_j$, @Poss$_j)$; and else (2) adding the axiom: Individual (φ (v_j) type (φ ($\lambda(v_j)$)) value (φ ($v_j$hasdp*PCDATA*), $\eta(v_i, v_j)$) [$\bowtie m_j$]), where $m_j = \eta(v_j$, @Poss$_j)$.

Here: (1) φ ($\lambda(v_i)$), φ ($\lambda(v_j)$), and φ ($\lambda(v_h)$) are fuzzy class identifiers $FCID_0$; φ ($v_i$hasopv_j) is a fuzzy object property identifier $FOPID_0$; and φ ($v_j$hasdp*PCDATA*) is a fuzzy datatype property identifier $FDPID_0$; (2) $\eta(v_i, v_j) = d_j \in$ **dom** is the content of the element $\lambda(v_j)$.

For example, for two nodes *university*, *UName* $\in N$ such that *university* < *UName* in the fuzzy XML document, we create the axioms: Individual (*id592914* Type (φ (*university*)) Value (φ (*university*hasop*UName*) *id593011*)); Individual (*id593011* Type (φ (*UName*))). Here, φ (*university*), φ (*UName*) are two fuzzy class identifiers $FCID_0$ and φ (*university*hasop*UName*) is the created fuzzy object property identifier $FOPID_0$, *id592914* and *id593011* are two fuzzy individual identifiers $FIID_0$ as mentioned in the rule 12. The examples for the other cases in the rule 13 can be given similarly.

Rule 14: For each pair nodes v_i, $v_j \in N$ with $v_i < v_j$, if $\lambda(v_i) \in \mathbf{E}$ and $\lambda(v_j) = @a_j \in \mathbf{A}$, then creating the axiom: Individual (φ (v_i) type (φ ($\lambda(v_i)$)) value (φ ($v_i$hasdpa_j), $\eta(v_i, v_j)$))).

Here: (1) φ ($\lambda(v_i)$) is a fuzzy class identifier $FCID_0$ and φ ($v_i$hasdpa_j) is a fuzzy datatype property identifier $FDPID_0$; (2) $\eta(v_i, v_j) = d_j \in$ **dom** is the value of the attribute. *For example*, for two nodes Val_1, $Poss_1 \in N$ such that $Val_1 < @Poss_1$ and $\eta(Val_1, @Poss_1) = 0.87$ in the fuzzy XML document, we create the axiom: Individual (*id593017* Type (φ (Val_1)) Value (φ (Val_1hasdp$Poss_1$) "0.87")). Here, φ (Val_1) is the fuzzy class identifier $FCID_0$ and φ (Val_1hasdp$Poss_1$) is the fuzzy datatype property identifier $FDPID_0$, and *id593017* is the fuzzy individual identifier $FIID_0$ created by the rule 12.Fmay be returned by invoking

Below we discuss the correctness of the approach. Being similar to the Theorem 7.1, the correctness can be sanctioned by establishing mappings between *instance documents* of the fuzzy DTD (i.e., *fuzzy XML documents*) and *models* of the constructed fuzzy ontology.

Theorem 7.6 *For every fuzzy DTD D and its transformed fuzzy ontology structure $\varphi(D)$, there exist two mappings* $\boldsymbol{\mu}$, *from fuzzy XML documents to models of $\varphi(D)$, and* $\boldsymbol{\lambda}$, *from models of $\varphi(D)$ to fuzzy XML documents, such that:*

- For each fuzzy XML document d conforming to D, $\mu(d)$ is a model of $\varphi(D)$;
- For each model FI of $\varphi(D)$, λ (FI) is a fuzzy XML document conforming to D.

The proof of Theorem 7.6, which is similar to the proof of Theorem 7.1, is omitted here.

In the following, we provide an example to explain well the transformation processes. From the approach above, the approach first performs the transformation from the symbols of fuzzy DTDs in fuzzy XML models (e.g., elements and attributes) to identifiers of fuzzy ontologies. For example:

1. The fuzzy XML element symbols such as *student*, *sname* in Fig. 3.2 are transformed into fuzzy class identifiers φ (*student*), φ (*sname*) $\in FCID_O$ respectively according to the rule 1.

   Then, the approach further transforms element type definitions of fuzzy DTDs in fuzzy XML models into fuzzy class and property axioms of fuzzy ontologies. For example:

2. For the element type definition such as *student* → (*sname*?, *age*?, *email*?) in Fig. 3.2, the following axioms are created by jointly using the Rules 6 and 11:

   Class (φ (*student*) complete intersectionOf (φ (*sname*) φ (*age*) φ (*email*)));
   Class (φ (*student*) partial restriction (φ (*studenthasopsname*) allValuesFrom (φ (*sname*)) maxCardinality (1)) restriction (φ (*studenthasopage*) allValuesFrom (φ (*age*)) maxCardinality (1)) restriction (φ (*studenthasopemail*) allValuesFrom (φ (*email*)) maxCardinality (1)));
   ObjectProperty (φ (*studenthasopsname*) domain (φ (*student*)) range (φ (*sname*)));
   ObjectProperty (φ (*studenthasopage*) domain (φ (*student*)) range (φ (*age*)));
   ObjectProperty (φ (*studenthasopemail*) domain (φ (*student*)) range (φ (*email*))).

   Finally, the approach transforms fuzzy XML documents of fuzzy XML models into fuzzy individual identifiers and axioms of fuzzy ontologies. For example:

3. For nodes *student*, *sname*, and *age* $\in N$ such that *student* < *sname*, *student* < *age* in the fuzzy XML document, the following axioms are created by jointly using the Rules 12 and 13:

Individual (*id593108* Type (φ (*student*)) Value (φ (*student*hasop*sname*) *id593216*) Value (φ (*student*hasop*age*) *id593237*));
Individual (*id593216* Type (φ (*sname*))); Individual (*id593237* Type (φ (*age*))).

The complete transformation example from a fuzzy XML model to a fuzzy ontology is provided in the following. Given a fragment of fuzzy DTD and fuzzy XML document (which are basically from the *fuzzy DTD* D_1 in Fig. 3.2 and the corresponding document instance *fuzzy XML document* d_1 in Fig. 3.1), according to the Rules 1–14, we can obtain the fuzzy ontology $FO = (FO_S, FO_I)$, which consists of the fuzzy ontology structure $FO_S = \varphi(D_1)$ and the fuzzy ontology instance $FO_I = \varphi\ (d_1)$ as follows. Here, some fuzzy axioms are omitted for brevity.

Class (φ(universities) partial restriction (φ(*universities*hasop*university*) allValuesFrom (φ(university))));

ObjectProperty (φ(*universities*hasop*university*) domain (φ(universities)) range (φ(university)));
Class (φ(university) complete intersectionOf (φ(UName) φ(Val_1)));
ObjectProperty (φ(*university*hasop*UName*) domain (φ(university)) range (φ(UName)) [Functional]);
ObjectProperty (φ(*university*hasopVal_1) domain (φ(university)) range (φ(Val_1)));
Class (φ(university) partial restriction (φ(*university*hasop*UName*) allValuesFrom (φ(UName)) Cardinality (1)) restriction (φ(*university*hasopVal_1) allValuesFrom (φ(Val_1)) minCardinality (1)));
Class (φ(UName) partial restriction (φ(*UName*hasdp*PCDATA*) allValuesFrom (φ(*PCDATA*)) cardinality (1)));
DatatypeProperty (φ(*UName*hasdp*PCDATA*) domain (φ(UName)) range (φ(*PCDATA*)) [Functional]);
Class (φ(Val_1) partial restriction (φ(Val_1hasop*student*) allValuesFrom (φ(student))));
ObjectProperty (φ(Val_1hasop*student*) domain (φ(Val_1)) range (φ(student)));
Class (φ(Val_1) partial restriction (φ(Val_1hasdp$Poss_1$) allValuesFrom (φ(CDATA)));
DatatypeProperty (φ(Val_1hasdp$Poss_1$) domain (φ(Val_1)) range(φ(CDATA)));
Class (φ(student) complete intersectionOf (φ(sname) φ(age) φ(email)));
Class (φ(student) partial restriction (φ(*student*hasop*sname*) allValuesFrom (φ(sname)) maxCardinality (1)) restriction (φ(*student*hasop*age*) allValuesFrom (φ(age)) maxCardinality (1)) restriction (φ(*student*hasop*email*) allValuesFrom (φ(email)) maxCardinality (1)));
ObjectProperty (φ(*student*hasop*sname*) domain (φ(student)) range (φ(sname)));
ObjectProperty (φ(*student*hasop*age*) domain (φ(student)) range (φ(age)));
ObjectProperty (φ(*student*hasop*email*) domain (φ(student)) range (φ(email)));
Class (φ(sname) partial restriction (φ(*sname*hasdp*PCDATA*) allValuesFrom (φ(*PCDATA*)) cardinality (1)));
DatatypeProperty (φ(*sname*hasdp*PCDATA*) domain (φ(sname)) range (φ(*PCDATA*)) [Functional]);
Class (φ(age) partial restriction (φ(*age*hasop$Dist_1$) allValuesFrom (φ($Dist_1$)) Cardinality (1)));
ObjectProperty (φ(*age*hasop$Dist_1$) domain (φ(age)) range (φ($Dist_1$)) [Functional]);
Class (φ($Dist_1$) partial restriction (φ($Dist_1$hasopVal_2) allValuesFrom (φ(Val_2)) minCardinality (1)));
ObjectProperty (φ($Dist_1$hasopVal_2) domain (φ($Dist_1$)) range (φ(Val_2)));
Class (φ($Dist_1$) partial restriction (φ($Dist_1$hasdp*type*) allValuesFrom (φ(*disjunctive*)));
DatatypeProperty (φ($Dist_1$hasdp*type*) domain (φ($Dist_1$)) range(φ(*disjunctive*)));
Class (φ(Val_2) partial restriction (φ(Val_2hasop*age_value*) allValuesFrom (φ(age_value)) Cardinality (1)));
ObjectProperty (φ(Val_2hasop*age_value*) domain (φ(Val_2)) range (φ(age_value)) [Functional]);
Individual (*id592846* Type (universities)

Value (*universities*hasop*university* id592912) [⋈ 0.87]
Value (*universities*hasop*university* id592914));
Individual (*id592914* Type (university)
Value (*university*hasop*UName* id593011));
Individual (*id593011* Type (UName)
Value (*UName*hasdp*PCDATA* "Wayne state University"));
Individual (*id592912* Type (university)
Value (*university*hasop*UName* id593015)
Value (*university*hasopVal_1 id593017));
Individual (*id593015* Type (UName)
Value (*UName*hasdp*PCDATA* "Oakland University"));
Individual (*id593017* Type (Val_1)
Value (Val_1hasop*student* id593108)
Value (Val_1hasdp$Poss_1$ "0.87"));
Individual (*id593108* Type (student)
Value (*student*hasop*sname* id593216)
Value (*student*hasop*age* id593237)
Value (*student*hasop*email* id593241));
Individual (*id593216* Type (sname)
Value (*sname*hasdp*PCDATA* "Tom Smith"));
Individual (*id593237* Type (age)
Value (*age*hasop$Dist_1$ id593319));
Individual (*id593407* Type (Val_2)
Value (Val_2hasop*age_value* id593516) [⋈ 0.9]
Value (Val_2hasdp$Poss_2$ "0.9")); …

So far, on the basis of the proposed approach, fuzzy ontologies can be constructed from fuzzy XML models. Two steps need to be carried out when transforming fuzzy XML model into fuzzy ontology: transforming the fuzzy DTD into a fuzzy ontology at structure level and transforming the fuzzy XML document w.r.t. the fuzzy DTD into the fuzzy ontology at instance level. Therefore, the correctness of the construction approach is decided by the respective correctness of two steps above. For the first step, the Rules 1–11 can first transform all the symbols in a fuzzy DTD (e.g., element names and attribute names) into fuzzy ontology identifiers, and then create a set of fuzzy class and property axioms from the element type definitions $E \rightarrow (\alpha, A)$ of the fuzzy DTD. The correctness of the transformation in this step can be guaranteed by the Theorem 7.6. For the second step, as mentioned in the rules 12–14, based on the constructed fuzzy ontology identifiers and axioms in the first step, the rules in the second step can further map the values of elements and attributes in a fuzzy XML document to fuzzy individual axioms. The correctness of the transformation of instance level can be ensured by the correctness of the first step.

Following the approach above, we developed a prototype transformation tool called *FXML2FOnto*, which can automatically construct fuzzy ontologies from fuzzy XML models. In the following, we briefly introduce the design and implementation of *FXML2FOnto*. The implementation of *FXML2FOnto* is based on Java 2 JDK 1.6 platform, and the Graphical User Interface is exploited by using the java.awt and javax.swing packages. The core of *FXML2FOnto* supports the transformations from a fuzzy DTD to fuzzy ontology structure as well as from a fuzzy XML document w.r.t. the fuzzy DTD to fuzzy ontology instance. Figure 7.3 shows the overall architecture of *FXML2FOnto*.

As shown in Fig. 7.3, *FXML2FOnto* includes four main modules: *input module*, *parsing module*, *transformation module*, and *output module*. The *input module* inputs the fuzzy DTD and the corresponding fuzzy XML document files, and it should be noted that at present *FXML2FOnto* cannot automatically extract the fuzzy DTD according to the fuzzy XML document if there is not a fuzzy DTD at hand; the *parsing module* uses Java methods to parse the fuzzy DTD and uses the XML DOM technique to parse the fuzzy XML document, and then stores the parsed results as document object trees; the *transformation module* uses XSLT processor to transform the parsed fuzzy DTD and fuzzy XML document into the fuzzy ontology structure and instance based on the proposed approaches above; the *output module* produces the resulting fuzzy ontology which is saved as a text file and displayed on the tool screen. The main class diagram of the tool is shown in Fig. 7.4.

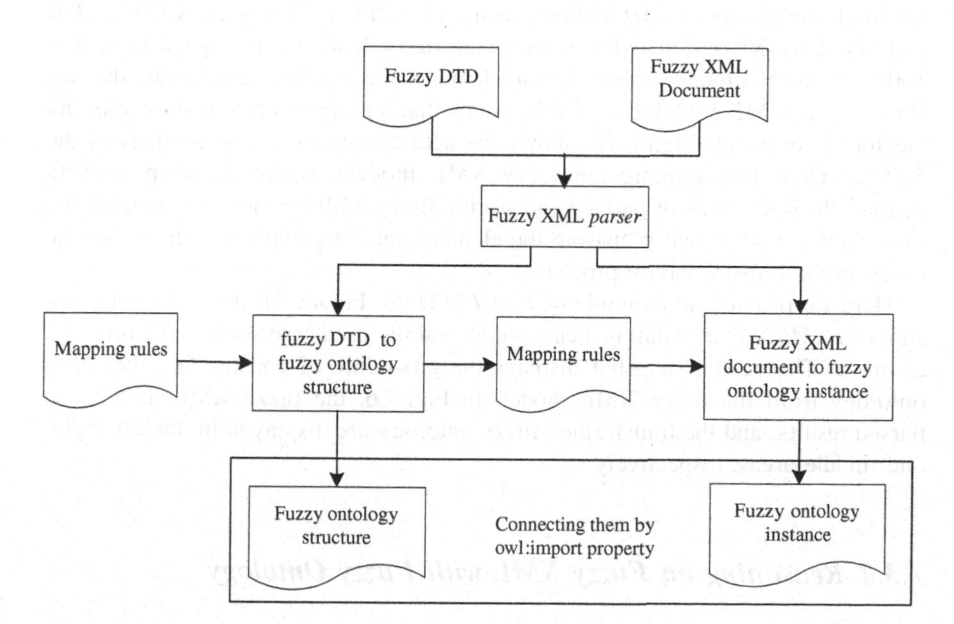

Fig. 7.3 The overall architecture of *FXML2FOnto*

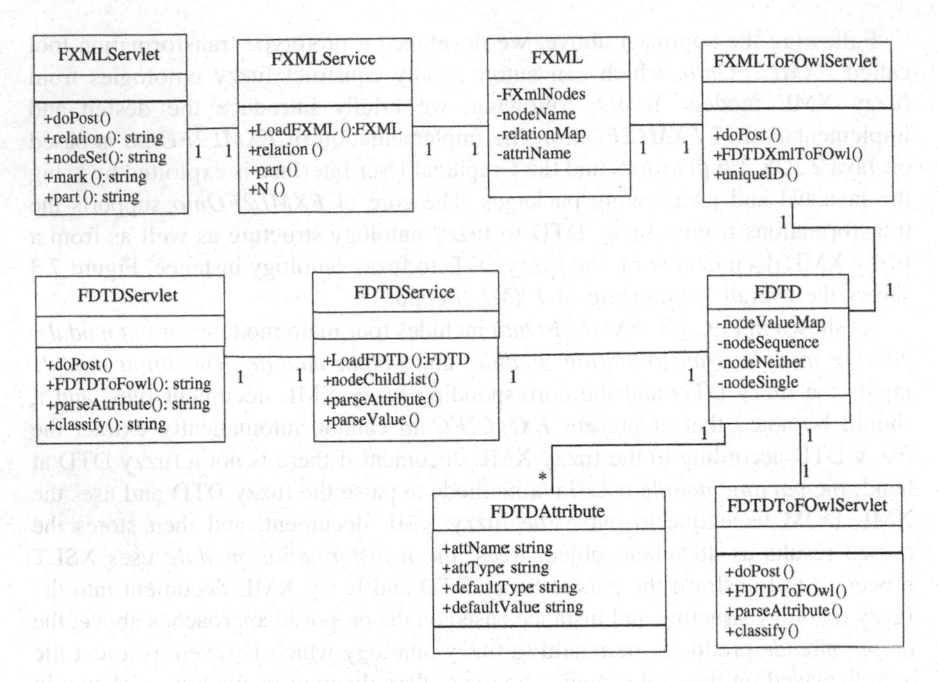

Fig. 7.4 The main class diagram of *FXML2FOnto*

We carried out transformation experiments of some fuzzy XML models using the implemented tool *FXML2FOnto*, with a PC (CPU P4/3.0 GHz, RAM 3.0 GB and Windows XP system). The sizes of the fuzzy XML models range from 0 to 1000. We use the total number of main elements and attributes to measure the size *N* of a fuzzy XML model. Case studies show that our approach actually works and the tool is efficient. Figure 7.5 shows the actual execution time routines in the *FXML2FOnto* tool running ten fuzzy XML models, where the preprocessing denotes the operations of parsing and storing fuzzy XML models, i.e., parsing the fuzzy XML models and preparing the element data in computer memory for the usage in the transformation procedure.

Here we provide an example of *FXML2FOnto*. Figure 7.6 is the screen snapshot of *FXML2FOnto* running a case study, and the used case study is the previous example. The screen snapshot displays the procedure of constructing the fuzzy ontology from the fuzzy XML model. In Fig. 7.6, the fuzzy XML model, the parsed results, and the transformed fuzzy ontology are displayed in the left, right, and middle areas, respectively.

7.3.2 Reasoning on Fuzzy XML with Fuzzy Ontology

Reasoning on fuzzy XML with fuzzy ontology is very similar to the reasoning of fuzzy XML with fuzzy Description Logic as introduced in Sect. 7.2.3. Fuzzy

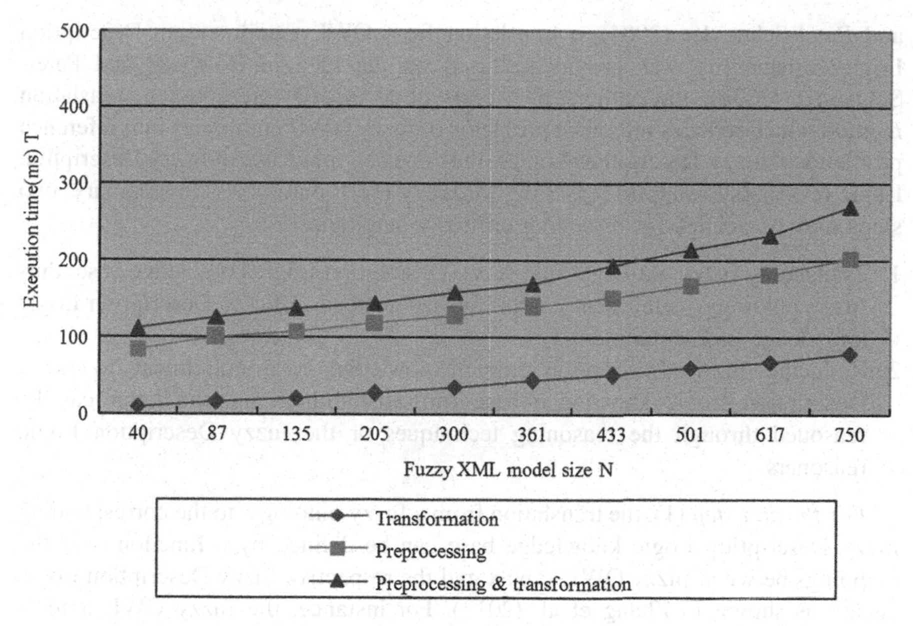

Fig. 7.5 The execution time of *FXML2FOnto* routine on several fuzzy XML models

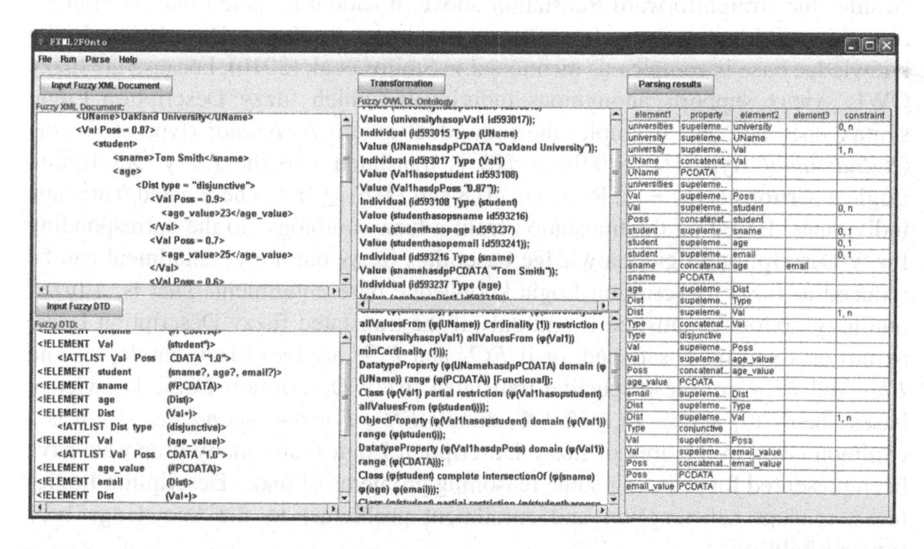

Fig. 7.6 The screen snapshot of *FXML2FOnto*

ontologies have the reasoning nature because of the existence of fuzzy Description Logics, since the logical underpinnings of fuzzy ontologies are mainly very expressive fuzzy Description Logic. Therefore, here we do not attempt to introduce the details of reasoning on fuzzy XML with fuzzy ontologies. In Horrocks and Patel-Schneider (2004), a translation from OWL entailment to Description Logic satisfiability was provided. Based on the idea in Horrocks and Patel-Schneider (2004), the authors in Stoilos et al. (2010) presented a translation method which reduces inference problems of fuzzy OWL ontologies into inference problems of fuzzy Description Logics, thus one can make use of fuzzy Description Logic reasoners to support reasoning for fuzzy OWL ontologies. In summary, two steps may be needed for reasoning on fuzzy ontologies:

1. Translating fuzzy ontology into fuzzy Description Logic knowledge base, thus fuzzy ontology entailment can be further reduced to fuzzy Description Logic knowledge base entailment.
2. Reducing fuzzy Description Logic knowledge base entailment to fuzzy Description Logic knowledge base (un)satisfiability, and the latter can be reasoned through the reasoning techniques or the fuzzy Description Logic reasoners.

For the first step (1), the translation from a fuzzy ontology to the corresponding fuzzy Description Logic knowledge base can be defined by a function over the mappings between fuzzy OWL syntax and the respective fuzzy Description Logic syntax as shown in Zhang et al. (2013). For instance, the fuzzy OWL axioms *SubClassOf* (C_1, C_2) and *EquivalentClasses* (C_1, …, C_n) are mapped to fuzzy Description Logic concept axioms $C_1 \sqsubseteq C_2$ and $C_1 \equiv \cdots \equiv C_n$ respectively. Besides the straightforward translation above, it should be noted that the translation of individual axioms from fuzzy ontology to fuzzy Description Logic knowledge base is complex as mentioned in Stoilos et al. (2010), because the fuzzy OWL syntax supports anonymous individuals, which fuzzy Description Logic syntax does not. For example, the individual axiom *Individual* (type (*C*) value (*R Individual* (type (*D*) $\geq$ 0.9)) > 0.8) is translated into the fuzzy Description Logic assertions $C(a) = 1$, $R(a, b) > 0.8$ and $D(b) \geq 0.9$, where *a* and *b* are new individuals. Based on the translation from a fuzzy ontology to the corresponding fuzzy Description Logic knowledge base, thus fuzzy ontology entailment can be reduced to fuzzy Description Logic knowledge base entailment. That is, a fuzzy ontology axiom is satisfied if and only if the translated fuzzy Description Logic axiom or assertion is satisfied, or if $F\mathcal{O}_1$ and $F\mathcal{O}_2$ are two fuzzy ontologies and $F\mathcal{K}_1$ and $F\mathcal{K}_2$ are the respective translated fuzzy Description Logic knowledge bases, then $F\mathcal{O}_1$ entails $F\mathcal{O}_2$ iff $F\mathcal{K}_1$ entails $F\mathcal{K}_2$. *For the second step* (2), it is a common issue in the area of fuzzy Description Logics, and many proposals have been presented for reducing some reasoning problems of fuzzy Description Logics (e.g., concept subsumption and entailment problems) to the knowledge base (un)satisfiability.

Overall, the translation approaches from fuzzy ontologies to fuzzy Description Logic knowledge bases together with the fuzzy Description Logic reasoning techniques as introduced briefly in Sect. 7.2, may provide reasoning support over fuzzy ontologies. In this section, the details of reasoning on fuzzy XML with fuzzy ontology, which are very similar to the reasoning of fuzzy XML with fuzzy Description Logic as introduced in Sect. 7.2.3, and thus are not introduced in here again. Based on the Theorems in Sect. 7.2.3, here we only give the brief reasoning algorithm for fuzzy XML models with the fuzzy ontology. The following algorithm describes the overall process of reasoning on fuzzy XML models with the fuzzy ontologies, and the algorithm includes two main steps: constructing fuzzy ontologies from fuzzy XML models and reducing the reasoning tasks of the fuzzy XML models to the reasoning problems of the constructed fuzzy ontologies. The former is based on some previous proposed transformation rules, and the latter is based on several reducing procedures as shown in Theorems 7.2–7.5.

Algorithm// The algorithm describes the reasoning process of fuzzy XML models with fuzzy ontologies

---

Input: R_t, i.e., any one of reasoning tasks of a fuzzy XML model (D, d) mentioned in Sect. 7.2.3, where D may be a fuzzy DTD or a set $\{D_1, D_2, \ldots\}$ of fuzzy DTDs, and d is a fuzzy XML document
Output: true/false
Steps:

1. Constructing a fuzzy ontology $FO = (\varphi\ (D),\ \varphi\ (d))$ from the fuzzy XML model (D, d) according to the transformation Rules 1–14:
   - For each element symbol $E \in E \rightarrow (\alpha, A)$ in D, creating a fuzzy class identifier $\varphi\ (E)$ as Rule 1;
   - For each type T (i.e., #PCDATA or CDATA) in D, creating a fuzzy data range identifier $\varphi\ (T)$ as Rule 2;
   - For each attribute type definition in D, creating the corresponding fuzzy ontology identifiers and axioms as Rule 3;
   - For each element type definition in D, creating the corresponding fuzzy ontology identifiers and axioms as the rules 4–11;
   - For each element node $v \in N$ in d, creating the corresponding fuzzy ontology individual identifier and individual axiom as Rule 12;
   - For each pair nodes v_i, $v_j \in N$ in d with $v_i < v_j$, creating the corresponding fuzzy ontology individual identifiers and individual axioms as the rules 13–14;
2. Reducing the input reasoning task R_t of the fuzzy XML model to the reasoning problem of the constructed fuzzy ontology FO:
   - If R_t is whether d conforms to D, then checking whether $\mu\ (d)$ is a model of $\varphi(D)$ as mentioned in Theorem 7.6, and then *goto* step 3;
     //$\mu\ (d)$ can be given according to Theorem 7.6
   - If R_t is $D_1 \subseteq D_2$, then checking whether $\varphi\ (D) \vDash r'_{D1} \sqsubseteq r''_{D2}$, and then *goto* step 3;
   - If R_t is $D_1 \equiv D_2$, then checking whether $\varphi\ (D) \vDash r'_{D1} \equiv r''_{D2}$, and then *goto* step 3;

- If R_t is $D_1 \otimes D_2$, then checking whether $\varphi(D) \models r'_{D1} \sqcap r''_{D2} \sqsubseteq \bot$, and then *goto* step 3;
  //here r'_{D1} and r''_{D2} are two fuzzy concepts in $\varphi(D)$

3. **Return** true/false
   //The reasoning results may be returned by invoking the reasoner *DeLorean* (Bobillo et al. 2012). When we input the constructed fuzzy ontology *FO* and the reasoning problems of *FO* above into the reasoner, the reasoner can return the results.
4. **End**

7.4 Summary

Data management involving XML with imprecision and uncertainty has attracted much attention both from academia and industry. Accordingly, a significant interest developed regarding the problem of describing fuzzy XML with expressive knowledge representation techniques in recent years, so that some fuzzy XML issues such as reasoning and querying may be handled intelligently. Reasoning on XML with imprecision and uncertainty would help to check whether a document conforms to a given document structure or two documents are compatible, and also may improve the precision and efficiency of query processing. Therefore, based on the high expressive power and effective reasoning services of the knowledge representation formalisms fuzzy Description Logics and fuzzy ontologies, in this chapter, we introduced how to reason on fuzzy XML with fuzzy Description Logics and fuzzy ontologies. We developed some rules for transforming fuzzy XML into fuzzy Description Logic and fuzzy ontology, respectively, and then presented approaches for reasoning on fuzzy XML with the transformed formalisms.

Reasoning on fuzzy XML is an important issue for establishing the overall management system and realizing the intelligent processing of fuzzy XML data. It should be noted that the issue of reasoning on fuzzy XML we introduced in this chapter mainly focuses on XML DTD because it has traditionally been the most common method for describing the structure of XML instance documents, and reasoning on fuzzy XML Schema can be done following the similar procedure.

References

Arenas M, Fan W, Libkin L (2002) On verifying consistency of XML specifications. In: Proceedings of PODS, pp 259–270

Buneman P, Davidson S, Fan W, Hara C, Tan W (2003) Reasoning about keys for XML. Inf Syst 28(8):1037–1063

Baader F, Calvanese D, McGuinness D, Nardi D, Patel-Schneider PF (2003) The description logic handbook: theory, implementation and applications. Cambridge University Press, Cambridge

Bray T, Paoli J, Sperberg-McQueen CM (1998) Extensible markup language (XML). W3C recommendation

Bobillo F (2008) Managing vagueness in ontologies. PhD Dissertation, University of Granada, Spain

Bobillo F, Delgado M, Gómez-Romero J (2012) DeLorean: a reasoner for fuzzy OWL 2. Expert Syst Appl 39(1):258–272

Borgida A (1995) Description logics in data management. IEEE Trans Knowl Data Eng (TKDE) 7(5):671–682

Bosc P, Kraft DH, Petry FE (2005) Fuzzy sets in database and information systems: Status and opportunities. Fuzzy Sets Syst 156(3):418–426

Calvanese D, De Giacomo G, Lenzerini M (1999) Representing and reasoning on XML documents: a description logic approach. J Logic Comput 9(3):295–318

Cautis B, Abiteboul S, Milo T (2007) Reasoning about XML update constraints. In: Proceedings of PODS, pp 195–204

Calegari S, Ciucci D (2007) Fuzzy ontology, fuzzy description logics and fuzzy-owl. In: Proceedings of WILF 2007, LNCS, vol 4578

Calvanese D, De Giacomo G, Lenzerini M (1998) What can knowledge representation do for semi-structured data? Proceedings of AAAI, pp 205–210

Drabent W, Wilk A (2006) Combining XML querying with ontology reasoning: Xcerpt and DIG. In: Proceedings of RuleML-06 workshop: ontology and rule integration. Athens, Georgia, USA

De Giacomo G, Lenzerini M (1994) Boosting the correspondence between description logics and propositional dynamic logics. In: Proceedings of AAAI, pp 205–212

Euzenat J (2001) Preserving modularity in XML encoding of description logics. In: Proceedings of description logic (DL 2001), pp 20–29

Galindo J (ed) (2008) Handbook of research on fuzzy information processing in databases. Information Science Reference, Hershey, pp 55–95

Horrocks I, Patel-Schneider PF (2004) Reducing OWL entailment to description logic satisfiability. J Web Semant 1(4):345–357

Lam THW (2006) Fuzzy ontology map-a fuzzy extension of the hard-constraint ontology. In: Proceedings of the 5th the IEEE/WIC/ACM international conference on web intelligence. Hong Kong, China, pp 506–509

Libkin L, Sirangelo C (2008) Reasoning about XML with temporal logics and automata. In: Proceedings of LPAR, pp 97–112

Lee CS, Jian ZW, Huang LK (2005) A fuzzy ontology and its application to news summarization. IEEE Trans Syst Man Cybern Part B 35(5):859–880

Lukasiewicz T, Straccia U (2008) Managing uncertainty and vagueness in description logics for the Semantic Web. Web Semant Sci Serv Agents World Wide Web 6:291–308

Ma ZM, Zhang F, Wang HL, Yan L (2013) An overview of fuzzy description logics for the semantic web. Knowl Eng Rev 28(1):1–34

OWL: Ontology web language. http://www.w3.org/2004/OWL/

Staab S, Studer R (2004) Handbook on ontologies. Springer, Berlin

Sanchez E, Yamanoi T (2006) Fuzzy ontologies for the semantic web. In: Proceedings of FQAS, pp 691–699

Stoilos G, Stamou G, Tzouvaras V, Pan JZ, Horrocks I (2005) The fuzzy description logic f-SHIN. In: Proceedings of the international workshop on uncertainty reasoning for the semantic web. CEUR-WS.org Publishers, Aachen, pp 67–76

Stoilos G, Stamou G, Kollias S (2008) Reasoning with qualified cardinality restrictions in fuzzy description logics. In: Proceedings of the 17th IEEE international conference on fuzzy systems (FUZZ-IEEE 2008). IEEE Computer Society, China, pp 637–644

Stoilos G, Stamou G, Pan JZ (2010) Fuzzy extensions of OWL: logical properties and reduction to fuzzy description logics. Int J Approximate Reasoning 51:656–679
Sánchez D, Tettamanzi A (2006) Fuzzy quantification in fuzzy description logics. In: Sanchez E (ed) Capturing intelligence: fuzzy logic and the semantic web. Elsevier, Amsterdam
Straccia U (2001) Reasoning within fuzzy description logics. J Artif Intell Res 14(1):137–166
Straccia U (1998) A fuzzy description logic. In: Proceedings of the 15th national conference on artificial intelligence, AAAI Press, pp 594–599
Straccia U (2005) Towards a fuzzy description logic for the semantic web. In: Proceedings of the 2nd European semantic web conference (ESWC 2005), pp 167–181
Tresp C, Molitor R (1998) A description logic for vague knowledge. In: Proceedings of the 13th European conference on artificial intelligence (ECAI-98), Brighton (England)
Toman D, Weddell G (2005) On reasoning about structural equality in XML: a description logic approach. Theoret Comput Sci 336(11):181–203
Wu X, Ratcliffe D, Cameron M (2008) XML Schema representation and reasoning: a description logic method. In: Proceedings of the 2008 IEEE congress on services, pp 487–494
Wood D (1995) Standard generalized markup language: mathematical and philosophical issues. Computer Science Today, pp 344–365
Yen J (1991) Generalizing term subsumption languages to fuzzy logic. In: Proceedings of the 12th international joint conference on artificial intelligence (IJCAI-91), pp 472–477
Zadeh LA (1965) Fuzzy sets. Inf Control 8(3):338–353
Zhang F, Yan L, Ma ZM, Cheng JW (2011a) Knowledge representation and reasoning of XML with ontology. In: Proceedings of the 2011 ACM symposium on applied computing, pp 1705–1710
Zhang F, Yan L, Ma ZM, Wang X (2011b) Representation and reasoning of fuzzy XML model with fuzzy description logic. Chin J Comput 34(8):1437–1451
Zhang F, Ma ZM, Yan L (2013) Construction of fuzzy ontologies from fuzzy XML models Knowl Based Syst 42:20–39

Index

L. Yan et al., *Fuzzy XML Data Management*, Studies in Fuzziness and Soft Computing 311, DOI: 10.1007/978-3-642-44899-7,

Printed in the United States
By Bookmasters